THE LIBRARY OF PRINTING TECHNOLOGY

MACHINE PRINTING

Other books in the Series

DESIGN FOR PRINT PRODUCTION
H S Warford

LETTER ASSEMBLY IN PRINTING
D Wooldridge

FINISHING PROCESSES IN PRINTING
A G Martin

MATERIALS IN PRINTING PROCESSES
L C Young

GRAPHIC REPRODUCTION PROCESSES
E Chambers

Series Editor
J E REEVE FOWKES MIOP
Head of the Department of Printing
Southampton College of Art

THE LIBRARY OF PRINTING TECHNOLOGY

Machine Printing

BY W R DURRANT, C W MEACOCK *and*
R E WHITWORTH

Focal Press, London and New York

ISBN 0 240 50821 1
Printed in Great Britain
by W & J Mackay Limited, Chatham

CONTENTS

Editor's Preface

Printing, in common with other great industries, is caught up in nuclear-age technology. For some printers this is a painful experience: the deep roots of five centuries of tradition resist the pull of technology. The pride of craftsmanship, with all its ingrained mystique, conflicts with the self-assurance of technological skills – skills which reduce the art and mystery of printing to binary numbers.

Already great changes have taken place within the printing industry: more are to come and the rate of change will accelerate. What is happening needs to be understood by printers if they are to remain masters of their destiny. More printers must be trained as technicians and technologists if the benefits of the 'knowledge explosion' are to be fully exploited.

Education and training for the printing industry is undergoing its own revolution to meet the challenge. Courses for printing technicians and technologists have been established: in Great Britain the Institute of Printing has fostered National Certificate, Diploma and Degree courses in printing technology, and the City & Guilds of London Institute offers examinations for a Printing Technician Certificate.

These new courses need books, and just as the industry is being overtaken by events, so, too, is the printer's library. This problem was seen by leading educationists among whom were Charles L Pickering, OBE, teacher of printing and for many years H.M. Inspector of Schools with special interest in printing education; H M Cartwright, former Principal of Bolt Court School of Photo-Engraving and a leading authority on photo-mechanical processes; and L T Owens, Principal of the London College of Printing and a member of the Printing and Publishing Industry Training Board. It is largely through their advice and help that this new Library of Printing Technology has come to fruition.

This series is unique in that it embraces, for the first time, the complete spectrum of printing technology: pre-production design and planning, letter assembly, graphic reproduction and printing surface preparation, machine printing, print finishing, and the technology of printing materials.

Editor's Preface

Machine printing, the subject of this book, depends to a very large extent on precision engineering of presses for successful transference of ink from an image surface to a substrate, and the impact of space-age technology is more evident in the machine manufacturing processes than in the machines themselves. However, few printers can be unaware of the changes which are taking place – the steady growth of offset lithography at the expense of letterpress, and the expansion of gravure to meet the demands of packaging. The traditional techniques of printing from relief surfaces are dying, but it would be rash to assume that offset lithography will be the major printing process of the future: high resolution electrostatic printing, or some other form of image transference from a film assembly rather than from a plate or cylinder may well oust all the major printing processes.

It is therefore important for tomorrow's craftsmen and technicians that they should have a knowledge of the principles and practice of all the printing techniques and not just experience in one narrow sector. With such knowledge they will be able to accept inevitable changes more readily, and exploit with confidence the machine printing processes of the future.

J E Reeve Fowkes

Acknowledgements

We are indebted to the following manufacturers for permission to reproduce illustrations and technical information:

Hawthorn Baker Limited
Crosfield Electronics Limited
Heidelberg Printing Machinery Company Limited
Mailänder Druckmaschinenfabrik
Mount Hope Machinery Limited
K. S. Paul (Koenig & Bauer) Limited
Timsons (Sales) Limited

1. Machine design

Printing is an abbreviation of the word imprinting, which in turn is synonymous with impressing. Printing machines are designed to impress ink onto the surface of a substrate such as paper, in selective areas, in order to inform or decorate. The essential difference between printing and dyeing or staining is that printing ink is intended to remain bonded at the surface of the substrate rather than penetrating into it. This situation is not always obtained, but one of the criteria of printing quality is the ability of the ink to 'sit up' on the surface.

FUNCTIONS OF A PRINTING MACHINE

The basic elements of a printing machine are: 1, a device for selecting the areas to be inked; this is the printing surface known as a forme or a plate; 2, a method of applying the ink to the printing surface; 3, a means of impressing the inked printing surface onto the substrate; and 4, a system for bringing the substrate to the printing position and then removing it. Each of these elements may vary in principle according to the design of the machine and the purpose for which it is intended.

The basic items will be contained in a common framework, and each one provided with means of adjustment. As the framework is tailored to accommodate the interior equipment a competent technician will recognise the purpose and capabilities of a machine from its outward appearance.

A control system is also an integral part of the machine. This is necessary to bring the separate parts into operation as required and may also co-ordinate the sequence of operations as well as controlling the adjustments and ancillary devices on the machine. Simple machines may require considerable operator involvement in the control system, but this would be considered poor design in a modern machine which should function efficiently with a minimum of operator interference.

Each of the above basic elements is treated separately in the following chapters, but it will be useful initially to cover briefly the fundamental considerations which influence printing machine design.

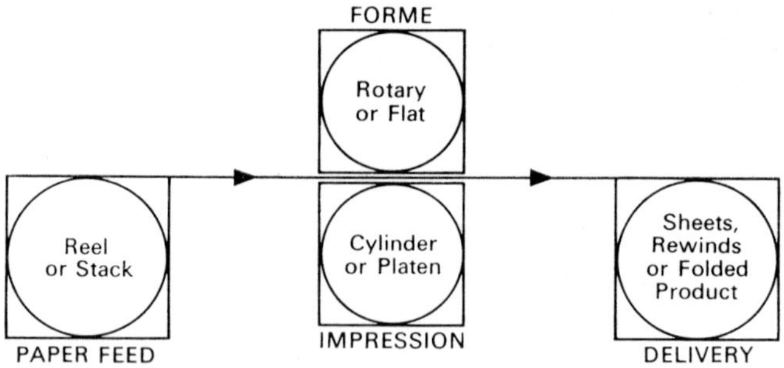

1.1 *Basic relationships of the main elements of a letterpress printing machine.*

Selecting the areas to be inked

The most common methods are:

1 To create a height differential in the printing surface so that the areas which are not required to transfer ink are below the rest; therefore ink applied to the surface is only laid on the raised image. This is the relief system generally referred to as letterpress, a variation of which is flexography (see appendix 8).

2 To obtain separation of printing and non-printing areas by chemical means on an otherwise plane surface so that the non-image areas repel ink. This is the planographic principle employed in lithography and collotype.

3 To arrange a height differential in the printing surface whereby the areas to be printed are recessed into it; the surface being flooded with ink and the surplus removed from the face so that ink is drawn from the recesses upon impression. This is the intaglio principle employed in photogravure.

4 To treat a permeable membrane so that only selected areas will allow ink to pass through, thus forming a stencil. This is the principle of screen printing which is dealt with separately in the appendix.

Preparation of the image for each job is a specialised process usually carried out in a different department and brought to the machine as a forme, plate or cylinder. A forme is only used with the letterpress system and is an assembly comprising a number of separate items which make up the printing image. For instance it is possible to combine type for text matter and blocks for illustrations in the same forme, the areas between printing elements being filled with spacing material. A forme is locked into the machine on a flat bed. A plate is a one-piece printing surface containing both printing and non-printing areas and may be used by all the major printing processes. Plates can be printed flat or made in curved or flexible form to go round a cylinder. Complete printing cylinders are used in letterpress and gravure to provide an accurate and continuous printing surface.

Inking

It will be seen in Chapter 3 that the design of an inking system will be influenced by a number of factors including the printing process, class of work, the substrate and probably most important of all the physical nature of the ink itself.

Most systems have three main areas of control: 1, a reservoir of ink maintained as a continuous supply source for the rest of the system; 2, a metering system required to supply a regulated flow of ink; and 3, application to the plate or forme.

In letterpress and lithography, roller trains are used between the metering system and application to break down the structure of the stiff-bodied inks which these systems use. Gravure and flexography which use fluid inks have no need for extended roller systems, and with gravure in particular precise metering is not obtained until the surface ink is removed, leaving specific quantities of ink in the recesses.

Impression

The act of impression is the moment of truth. If there is any fault in the printing surface, the inking or the paper control, it will be evident when the inked image is impressed on the paper.

As with printing surface carriers the impressing surface may be either flat (a platen) or cylindrical. A platen is used only in conjunction with flat formes as there is no mechanical or other advantage in impressing a flat surface against a curved printing surface. Cylinder impression, on the other hand, may be used either with flat beds or in conjunction with plate cylinders to provide a complete rotary system.

Control of the substrate

The paper or other material may be fed to the printing unit either in separate sheets from a stack or continuously, as a web, from a reel. A possible alternative is to start with a reel of paper which is cut into sheets just before printing. Positive control must be maintained on the material at all times, whether it be in sheet form or web. Lack of adequate control will mean that the machine may function below its maximum efficiency; printing quality may suffer due to image slur, misregister or creasing of the stock; and wastage may be unnecessarily high.

Having passed through the printing unit the substrate must be further controlled to ensure that it is not creased or otherwise damaged and that the inked surface is not marked by other parts of the machine or by subsequent operations. The ideal requirements for sheet control and web control are fundamentally different and are covered in Chapters 5 and 6 respectively.

MACHINE CONSTRUCTION

It will be evident that the design of the items comprising a printing machine, as outlined at the beginning of this chapter, will be affected by a variety of factors, not the least being the type of work for which the machine is intended. Some of the major points to be considered are outlined in the following examples.

Letterpress machines

Sheet-fed platens are generally small machines taking a sheet up to A3 size and intended for general jobbing work. Compactness of design is usually an important feature as small firms may have space problems. This requirement implies close proximity of the component parts of the machine, plus the shortest paper travel line compatible with accessibility requirements. Versatility is also essential as the jobbing printer may only have one type of machine, and this may need to cope with a variety of work including numbering, embossing, perforating, scoring, cutting and creasing.

The construction of a modern platen requires a firm base and a rigid frame so that vibration does not occur when the platen is opening and closing at maximum speed. The platen may be capable of exerting a pressure of 5/MN m² on the forme, at speeds of around 5 000 impressions per hour. The outer edge of the platen is accelerated to over 650 mm per second before being stopped for the return movement.

The forme is held vertically in the machine and the forme rollers receive ink from a drum above the forme, rolling downwards to ink it, and returning to clear the platen nip, so that the forme is, in fact, rolled twice for each print. Before the impression is made, a sheet is taken from a feed pile at one side of the machine and placed in position on the platen. After the impression the sheet is taken from the platen and placed on the delivery pile.

Provision must be made for adjustment to the four basic mechanisms previously mentioned, i.e. forme, paper, ink and impression. This requirement is common to all machines, and each item is dealt with in more detail in the following chapters.

The great forces required for impression mechanisms preclude the general use of platens for larger sheet sizes. Some larger platens are made, but as a rule their speed decreases in proportion to the forme area. Their maximum efficiency is achieved in sizes where the impressional area is less than one tenth of a square metre, although even here a force of up to 50 tonnes may be produced.

The total force required to make a satisfactory print may be reduced by substituting rolling pressure for overall pressure. The theory of impression is developed further in Chapter 4, but in terms of machine design this means that larger flatbed machines will be designed with cylinders.

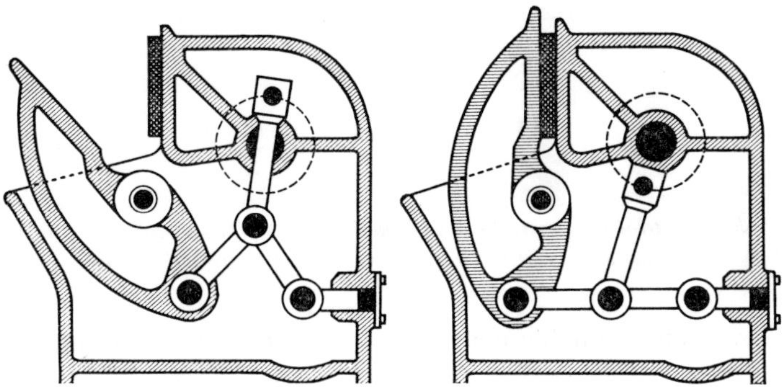

1.2 *Platen machine, showing the toggle lever action commonly used for opening and closing the platen.*

The vertical job press is an example of this principle, where the platen is replaced by a cylinder which rolls over the forme. It will be seen that this method of impression fundamentally affects the design of the machine. As the cylinder must clear the forme at both ends, the cylinder movement is reduced by reciprocating the bed in unison with the cylinder. Paper control is also entirely different as the sheets can no longer pass from side to side across the forme but must pass from front to back of the machine, around the cylinder.

The size limitations of this type of machine are largely associated with the handling of the forme. As the forme, in its chase, is placed manually in the machine about chest high, the weight of the forme is of prime consideration. The relative density of type metal is approximately 9·78 and this means that a forme of type completely filling the machine chase may weigh up to 40 kg. There is also the fact that a large forme is liable to distortion due to high internal stresses which are introduced when the component parts are locked into the chase. In order to lift a forme in which the parts are held together by lateral friction only, the forces exterted by the locking media, usually expanding quoins, must be greater than the total weight of the forme and chase. In

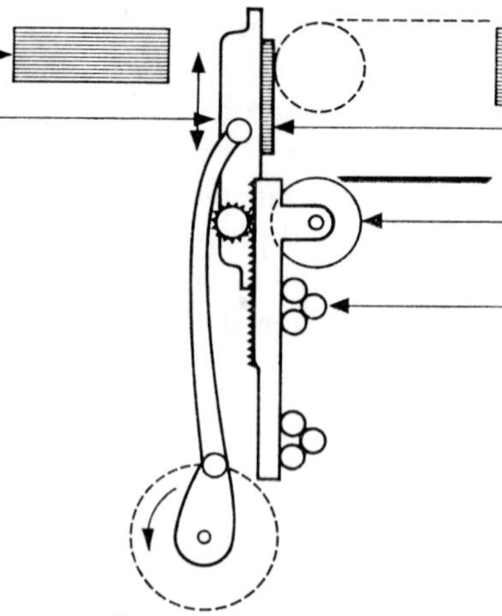

1.3 *Principle of vertical job press. The bed is designed to reciprocate, thus cutting down the overall movement of the cylinder as it rolls across the forme.*

large formes this inevitably means distortion of the printing image and the chase.

Cylinder machines are usually constructed with horizontal beds for the above reasons and others, associated with forme handling and paper control. The forme may slide onto the bed from a trolley, thus minimising the risk of the forme falling apart during handling. The machine printer may also safely reduce the pressure on the quoins and 'plane' the forme on the machine to ensure that it is lying flat on the bed and without internal distortion.

The smaller cylinder machines (up to A2 sheet size) are usually constructed as compact, one-piece units. The dimensions of the machine are generally no more than four times the dimensions of its maximum sheet size (420mm × 594mm) and may not occupy an area much more than, say 1·6m × 2·4m. A well-designed machine will have complete accessibility of all working areas with the fewest possible number of operating positions, and without climbing or reaching. Another feature of good design is to have the minimum number of machine adjustments compatible with adequate control.

Most small cylinder machines of modern design have certain common features in their basic construction. The impression cylinder is centred on the bed travel and paper control systems are positioned at one end of the machine in order to give open access to the forme and inking system at the opposite end. Paper is fed to the impression cylinder via a feed-board positioned above the cylinder. The sheets are taken around the cylinder to make the print and from there to the delivery pile below the feeder.

For operating convenience and efficiency no part of the machine should be above eye level, the highest point usually being the top of the feeder pile with its sheet separating mechanism. It will be evident that the depth of the feeder pile is therefore limited to allow for the sheet delivery system and the delivery pile itself. In practice the feeder pile rarely exceeds more than 30 per cent of the total height of the machine and higher piles generally mean loading (by hand) above eye level. A pile of 500 mm depth will contain approximately 5 000 sheets of average caliper stock. This may well be satisfactory for general jobbing work where runs rarely amount to tens of thousands, but for extended running it may be necessary to incorporate an automatic pre-loading device so that the machine may be run continuously.

A number of basic design variables on small cylinder machines are concerned with paper control. For instance, to ensure correct positioning of the sheet to the impression cylinder it must be located to three accurately-set positions, or lays, at the end of the feed board; two lays for the front edge of the sheet, and one for the side. On older types of machines the sheet was snatched directly from the feedboard by grippers in the cylinder, and a high degree of skill was required of the machine printer to ensure that the timing of the gripper action coincided precisely with the movement of the lays. Snatching too soon meant pulling the sheet before the lays released it, while taking the sheet too late meant that it was temporarily uncontrolled.

Stop cylinder machines. A modern machine will control the feedboard-to-cylinder action in one of two basic ways. The first is to stop the cylinder rotation temporarily while the sheet transfer action takes place. The 'stop-cylinder' system allows the relationship of cylinder and feedboard to be accurately determined and the cylinder grippers may close upon the sheet before the lays release it, ensuring that the paper is never out of positive control. The stationary period of the cylinder is also used to allow the bed to return after its printing stroke to the pre-printing position.

Single-revolution machines. The second method of feedboard-to-cylinder control is the transfer gripper system. This is a set of intermediate grippers pivoted on a line equidistant from the feedboard lay position and the cylinder periphery. The transfer gripper arm often contains the front lays so that the sheet coming down the feedboard is positioned to the lays while the transfer grippers are in a stationary position at the feedboard. The transfer arm then accelerates the sheet to the cylinder where a smooth, synchronised changeover takes place.

A primary advantage of the transfer gripper system is that it eliminates the need for stopping and starting mechanisms which are necessary on a stop-cylinder machine. However, as the cylinder revolves continuously, a bed return movement must occur while the cylinder is still rotating, and if a sheet is to be printed at every revolution of the cylinder then it follows that its circumference will be much larger than that of a stop-cylinder machine. This is not necessarily a disadvantage from the printing point of view, but is an important factor where machine design is concerned as the cylinder diameter on a single-revolution machine may account for more than 40 per cent of its total height.

The method of delivering the printed sheets on small cylinder machines will also vary. At the moment of impression the sheet will have its printed side downwards, and will need to be turned over before reaching the delivery pile so that the printer can examine the work while the machine is running. The inked side of the paper must be supported during turnover and unless proper provision is made there is a danger of marking the wet print. On some machines the sheet is supported on discs or serrated wheels which can be adjusted to the print margins; on other machines it is controlled by air blast to prevent the inked sheet touching the delivery mechanisms.

The stop-cylinder and the single revolution principles both have limitations where larger sheets are to be printed. A stop-cylinder to take a sheet size AO would require a cylinder at least 1500 mm wide × 350 mm diameter which would weigh about 1000 kg. Continual starting and stopping of such a

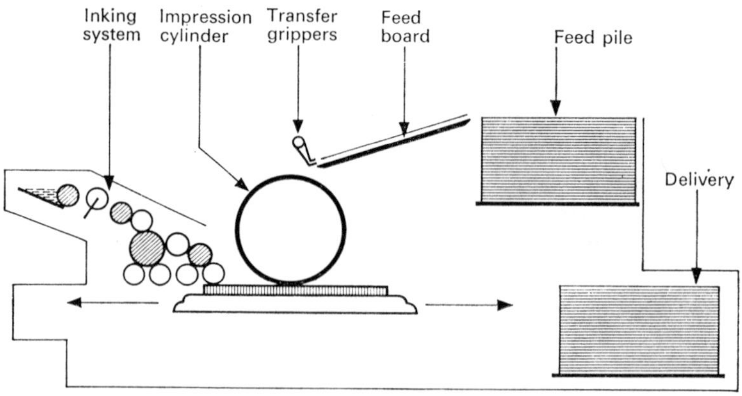

1.4 *Principle of single-revolution machine. The cylinder makes one complete revolution while the bed makes a double stroke for the printing and non-printing movements. Sheets of paper are transferred smoothly from a stationary position at the feedboard to the moving cylinder by means of swinging grippers.*

cylinder would use a great deal of power and require a particularly robust construction. A single-revolution machine for this size of sheet would require a cylinder nearly three times this diameter, which would pose further design problems. Paper handling would also be more difficult if the larger machines were constructed on the same lines due to the fact that the sheets must be lifted manually up to the feeder pile position.

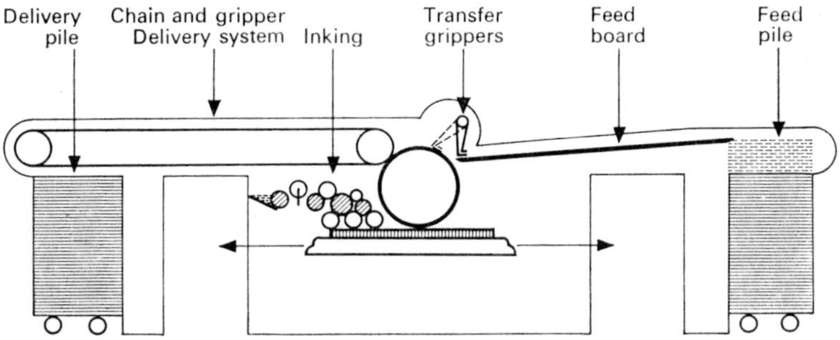

1.5 *Principle of two-revolution machine. The cylinder makes one revolution for the printing stroke of the bed and another revolution while the bed is returning. Sheets are delivered from the top of the cylinder and do not need to be turned over for delivery.*

Two-revolution machines are most commonly used for larger formes. The cylinder on this type of machine revolves continuously above the bed, but has a circumference only fractionally larger than the maximum sheet length. This means that the cylinder revolves once for each stroke of the bed, or two revolutions per print. On the printing stroke the cylinder bearers will be engaged in firm contact with the bed bearers, but will rise sufficiently to clear the forme on the return stroke. Transfer grippers will take the sheet from the feedboard to the impression cylinder.

It is common practice to arrange the feeder and delivery piles at opposite ends of the bed. This means that for maximum accessibility there must be a space between the piles and the bed, and this will extend the length of the machine considerably. This arrangement, however, allows a much deeper pile to be provided, up to the full height of the machine from floor level. A further advantage is the ability to truck in full pallets of paper, checked, 'knocked up' and ready for printing, and to remove stacks equally readily.

The sheets may be delivered from the top of the cylinder on the non-printing stroke, and as the paper is then print-side uppermost the problem of sheet turnover does not exist.

The speed of any flatbed machine is limited by the inertia of the bed and its forme. If the impulse of the force resulting from the change of direction of the bed is greater than the forces which hold the rest of the machine to the floor, then the machine will be unstable. Even if the frame is firmly anchored, undesirable stresses will occur within the machine.

It can be seen that the highest speeds will be obtained with beds of minimum length to reduce the time taken to complete the stroke, and minimum weight to reduce the inertia. The smaller cylinder machines may achieve speeds of 10 000 strokes per hour, which may be equivalent to the velocity of approximately 1·5 metres per second. Larger machines must make fewer strokes per hour at reduced velocities if they are to remain stable. Larger machines will therefore not necessarily bring a proportional increase in production.

Rotary machines. Where faster production is required and larger sheets are to be printed, the machine designer will employ the rotary principle. This requires a fundamentally different concept of the printing process as the bed is replaced by a cylinder and the flat forme by a curved plate or set of plates specially made to fit the cylinder. Providing that the forme requirements can be satisfactorily met, the rotary principle brings a number of additional benefits: for instance less power per equivalent mass of machinery is required. Both types of machine require a minimum of 1 watt per kilogramme to achieve an acceleration of one metre per second, but on a rotary machine most of the power is required to overcome the initial inertia of the cylinders when at rest, with a reduction in power once the cylinders are revolving. On a flatbed machine the inertia must be overcome at each stroke. Thus rotary machines may run at higher speeds with smaller motors than flatbeds of equivalent size.

The rotary principle also offers greater versatility of design. The plate and impression cylinders may be positioned horizontally, vertically or at any intermediate angle to each other as the design requirements indicate. Space economy may be made as units can be superimposed, and cylinders take less room than beds.

Better mechanical efficiency is also offered by the rotary principle. A sheet-fed rotary producing 6 000 sheets per hour will be doing less work than a flatbed machine of the two-revolution type running at 3 000 sheets per hour. The number of cylinder revolutions per hour will be identical, but the flatbed machine will require more power.

In rotary presses the units may easily be coupled. The rotary principle lends itself readily to the construction of machines with multiple units, without loss of mechanical speed.

Single-colour machines are used for various types of work in the medium range of runs from tens to hundreds of thousands. Their design follows a standard pattern of printing unit arranged vertically, impression cylinder below, plate cylinder centre and inking system above. Paper is fed into the nip between plate and impression cylinders via transfer grippers and must be firmly controlled until printing is complete. This control is usually obtained by retaining the sheet in the impression cylinder grippers during the printing part of the machine cycle.

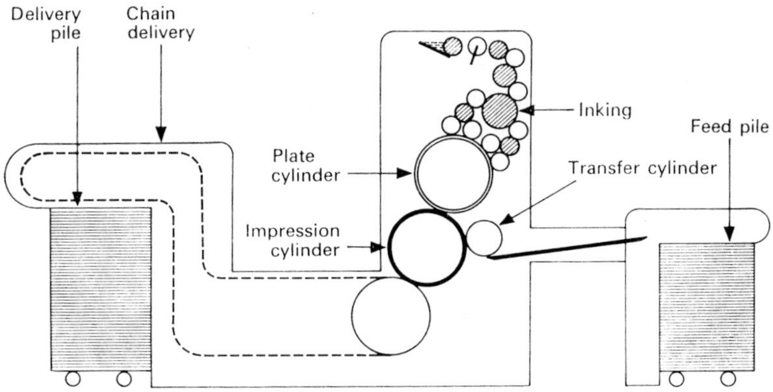

1.6 *Principle of a letterpress sheet-fed rotary machine. The greater efficiency of the rotary principle means that this type of machine will work faster, with less power and require less space than a flatbed printing a similar size of sheet. However, the advantages of the flat forme are lost as the machine must use rotary plates.*

The feeder and delivery piles on small machines may both be at the same end of the machine, allowing complete accessibility to the printing cylinders. This is ideal for small machines but not practical for larger sheets of paper and where deep piles are needed. Consequently most sheet-fed rotaries have feeder and delivery at opposite ends for convenience of loading and unloading, and the paper-path is diverted after printing to provide convenient access to plate and impression cylinders.

Composite machines. Where a printing firm has a regular supply of work which requires more than one colour to be printed on every sheet, then there are obvious practical advantages to be gained by employing machines with more than one printing unit. A four-colour catalogue can be printed quickly and economically on a four-colour machine, but it should be borne in mind that this machine is not practical for other types of work such as two-colour labels or bookwork. In other words the gain in productivity means a loss in versatility. For these reasons such machines are usually found in specialist houses or larger concerns which have sufficient volume of the specific type of work to keep the machines fully loaded.

There are two basic ways by which printing operations may be multiplied within a single machine. One method is to build in additional printing units similar to those on a single-colour machine so that printing becomes a repetitive process. Some machine manufacturers offer their customers the facility of building up machines from one to any number of colours in this manner. From the designer's point of view this system offers few problems as the only new mechanism required is sheet transport between units. The efficiency of this type of machine may be improved by slightly redesigning the units to achieve maximum accessibility and the minimum number of working positions.

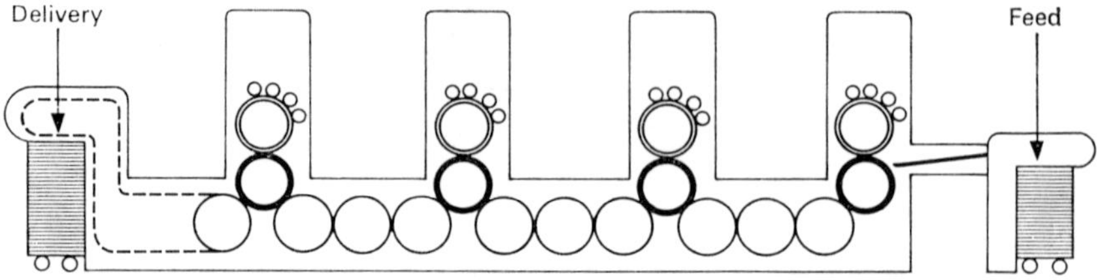

1.7 *Multicolour sheet-fed rotary, showing how printing units may be coupled together for multicolour printing. The sheets are carried from one unit to the next by grippers in the transfer cylinders between the units.*

The second method of constructing a multi colour machine is to group a number of plate cylinders, with their inking systems, around a common impression cylinder. This means a complete departure from the single-colour concept and results in a much more compact machine than one with separate units. There are also fewer moving parts and less sheet transport problems. The saving in floor space on this type of machine usually means some loss of accessibility, and this may require some parts of the machine to be constructed to swing or rack away, providing access to essential areas.

A compromise between these two types of multi colour machines is sometimes preferred. For instance a four-colour machine may be designed so that it has two printing units, each comprising a common impression cylinder with two plate cylinders. This system still keeps the sheet travel reasonably short, while avoiding some of the problems which may arise when overprinting four inks almost simultaneously.

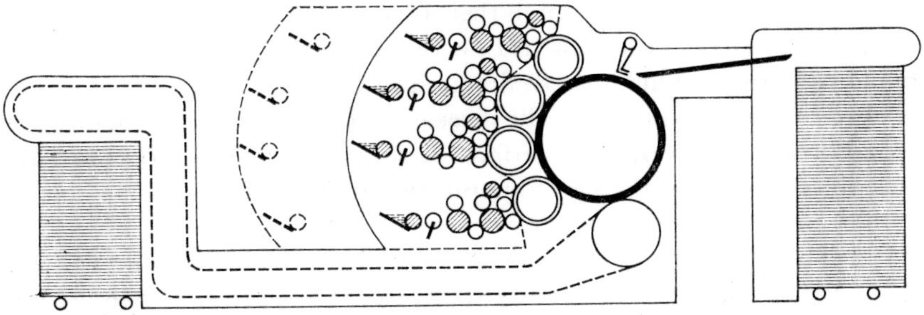

1.8 *Multicolour sheet-fed rotary, showing how plate cylinders may be grouped around one common impression cylinder. The inking systems can be moved away from the plate cylinders for accessibility.*

Perfectors. When a sheet of paper has been printed on both sides it is said to be 'perfected', and a machine which prints both sides of a sheet or web in one pass is called a perfector. Such a machine must necessarily have two inking systems, two plate cylinders and two impression cylinders. These may be arranged in the machine as two separate printing units with a sheet transfer system between them or the whole assembly may be contained in one framework, in which case it becomes an integral perfecting unit.

The printing units on a perfector are designated 'first couple' and 'second

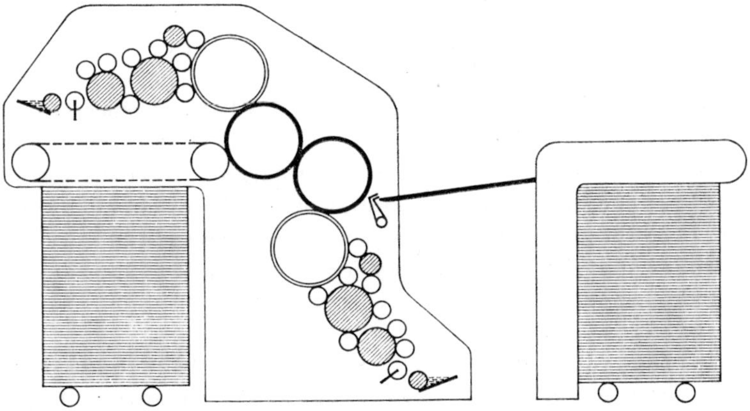

1.9 *Sheet-fed rotary perfector. The sheets are transferred directly from one impression cylinder to the next. Care must be taken to ensure that the wet ink from the first impression does not cause marking problems on the second impression.*

couple' respectively, and although their functions appear to be basically the same, there is a significant difference between the two when in operation. It should be borne in mind that the first unit receives a clean sheet and is only required to impress its ink onto the sheet against the impression cylinder. The second printer however, must perfect the sheet and in doing so presses the first printed side onto the second impression cylinder, usually while the ink is still wet.

It will be evident that there is some possibility of marking the first printed side if the wet ink should transfer to the second impression cylinder and thence onto subsequent sheets. This fault can be reduced by extending the sheet travel between couples so that the ink has more time to set, or by incorporating anti-set-off devices in the machine. The subject of 'second impression cylinder set-off' is dealt with more fully in Chapter 5.

Most perfectors are designed to print in black only. Their primary use is for book printing and similar publication work where the printing is mainly text matter. Multi colour perfectors are very seldom required by sheet-fed printers, but reel-fed machines printing several colours on both sides of a web are used for mass production of colour magazines and catalogues.

It should be noted that sheet-fed rotary machines do not run proportionately slower with larger sheet sizes as happens with flatbed machines. The maximum practical size of sheet with rotary machines is more likely to be determined by other factors such as width of doors or gangways, or the limitations of the bindery.

The two main factors which limit the speed of a sheet-fed rotary machine are 1, how quickly the sheets can be separated from each other and fed accurately to the printing unit, and 2, how efficiently the sheets can be stacked in the delivery. The printing operation can in fact be done much faster than either of these two functions.

The nature of the printing stock will considerably affect the speed of a sheet-fed machine, but even with an ideal paper the maximum speeds are around 10000 sheets per hour, or approximately one-third of a second per sheet. This is equivalent to an average lineal paper speed of about $2\frac{1}{2}$ metres per second when feeding sheets of AO size.

Web-fed machines. Much higher speeds are made possible by reel feeding, because the sheet-separating mechanism is no longer required, and in some respects paper handling problems are simplified. Therefore where the production of sheet-fed machines is insufficient for the volume of work required the designer must consider web feeding. Under favourable conditions paper speeds of more than 10 metres per second may be obtained.

However, the higher speeds obtained are not the only reasons for employing web feed. For instance, where the end product has a repetitive design, such as wallpaper, or must remain intact as with continuous stationery, then web feeding is essential. Similarly, certain materials such as plastic films and metallic foils in thin gauges are difficult to handle in sheet form, but comparatively easy to handle as webs. Machines for printing on such materials are invariably reel-fed.

A further advantage of web feeding is the ability to include multiple operations in addition to printing. In this way it is possible to perform many of the functions which in sheet printing would be done in the print finishing department. The result is that not only is printing much faster on web machines, but post-printing operations may be performed simultaneously, so that the end product of the machine may be a complete article requiring little or no further work to be done.

It follows that reel-fed machines are highly productive and particularly well suited to the production at high speeds of large quantities of printed products. They are most efficient when producing specialised types of work or a standardised range of products. In practice many reel-fed machines will only print a particular class of work, and as a rule the more versatile the machine, the lower its productivity. A completely versatile machine, designed to tackle all types of work would not be a practical proposition due to the quantity and variety of equipment that would need to be included to allow for all possible operations, most of which would not be in use for any one specific job.

A reel-fed machine then, will be a collection of printing units and ancillary equipment designed to produce a particular type of work. For instance, a newspaper machine will comprise a number of perfecting units, each fed with a separate web and all printing their own sections of the paper. The webs will be gathered, cut, folded and delivered at the end of the machine as complete newspapers. Packaging work, which is often multi-coloured but usually one-sided, may be printed on a unit with a number of plate cylinders grouped around a common impression cylinder to ensure colour register and control of difficult printing materials.

A machine for printing stationery will be different again, possibly comprising a number of single-colour printing units in line with a range of equipment for ancillary operations such as hole punching, perforating, slitting and numbering.

Some of the more common applications of web printing are dealt with in Chapter 7.

PRINCIPLES AND METHODS OF OFFSET PRINTING

The image from the printing surface is not always impressed directly onto the substrate. An intermediate cylinder with a resilient covering is sometimes used; the image being impressed upon this cylinder and then transferred to the substrate. The most common material for covering the offset cylinder is rubber, which has the useful ability to receive ink readily and also to lay it evenly on most substrates.

For letterpress printing offsetting the image has the advantage of greatly reduced impressional requirements, elimination of makeready, minimum plate wear and no indentation into the substrate. The gravure process uses the offset principle for specialised applications where it is often helpful in avoiding ink transfer problems.

Lithographic principles

Lithographic machines employ the offset principle extensively, to the degree that direct printing is the exception and is only used for specialist purposes. The reasons for this fact and the influence of the offset principle on machine design must be understood.

The basic principles which influence offset lithographic machine design can be stated under the following headings:

The printing cycle. A machine is designed to automate the hand production method. With manual lithography the prepared surface is first dampened to make the non-image areas ink repellent, then inked and finally impressed. This sequence must be maintained in machine printing and imposes certain restrictions on machine design and limitations on the ways in which space can be allocated to the various functions.

Control of ink/water balance. Although the printing surface is selective, the application of ink and water must be controlled within specific limits. Correct balance is maintained on modern machines by precision monitoring methods which hold the media within the prescribed tolerances.

Double image reversal. The main reason for using an intermediate cylinder in lithography is to obtain good control between the substrate and the inked surface without the high impressional stress that would be necessary with direct printing. The offset principle also enables a right-reading image to be used on the printing surface which can be an advantage in the plate preparation system. However, there is a loss of ink film thickness at impression due to the double transfer and this reduces the maximum film weight to about half of that obtainable by letterpress and one-sixth of the gravure potential.

Engineering requirements. The substrate is subjected to dimensional strain in offset printing due to the overall contact pressure in the printing nip. This has a considerable influence on the drive of the sheet or web as it passes

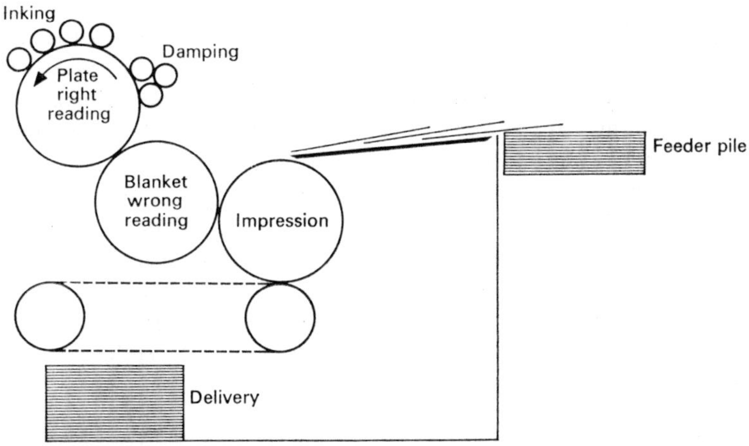

1.10 ***Principles of offset lithographic printing shown by this typical arrangement of machine elements. The sequence of events in the printing cycle are significant in all lithographic machines.***

through the machine. Hydrous stock may also be structurally affected by press moisture. For these reasons, feeding, transport and delivery mechanisms must be designed to cope with the above situations.

Interruptions to the sheet feeding will affect the ink/water balance with a risk of variations in the printed result. Complicated tripping arrangements are usually built in to reduce spoilage and to maintain correct printing balance.

Single-colour lithographic machines

The extra cylinder required for offset printing with its resilient surface and need for fine control of impression requires a system for lowering and lifting cylinders in and out of pressure contact when the feed of sheets is interrupted. If this provision were not made a tacky line would result at the place where the cylinders were left in contact; the printing plate would sensitise at this line and become receptive to ink, while the printing blanket would carry the tacky line to the next impression and the impression cylinder would mark the back of subsequent sheets.

To avoid these hazards the cylinder must move out of contact quickly in response to a signal from the feeder. The most common method is to turn the blanket cylinder in an eccentric bearing within two definite limits. These two positions should be finely adjustable and accurately reproducible. The linkage between feeder signal and eccentric bearing may be mechanical, electrical, electronic or hydraulic.

The plate cylinder must be accurately positioned in relation to the blanket cylinder, the inking rollers and the damping system. The pressures between these surfaces are critical and should be measurable and adjustable. It is equally important that the peripheral speeds of all rolling surfaces should be controllable within specified tolerances. This cylinder is therefore undercut to provide variable packing facilities under the plate to give limited control of surface speed. The surface contact line between cylinders is related to the circular pitch of the gears. Machine manuals specify the packing required to achieve this 'kiss' impression plus a recommended extra squeeze to gain satisfactory image transfer. This 'interference pressure' is in excess of normal packing to the pitch circle and is subject to adjustment depending on cylinder surfaces and printing conditions.

The blanket cylinder must also have provision for packing adjustment and the recess is cut to allow for one or two blankets to be used. A single blanket system will provide greater accuracy, while two blankets give a softer impression which is an advantage with solid work. Tensioning of the flexible blankets around the cylinder needs strict control and precaution against slippage; small machines usually tension at the back edge of the blanket while larger machines also tension from the front edge to allow for overall movement of the blanket so that scars on the surface will not coincide with the printing image. A method of paralleling the blanket cylinder to contacting surfaces is essential to control sheet drive and printing quality.

The impression cylinder has a hardened steel surface and requires no covering unless the machine is constructed on a bearer contact principle, in which case changes in sheet caliper require compensating adjustment to impression cylinder packing so that the substrate is kept in a positive relationship to the circular pitch of the cylinder drive. The disadvantage here is an increase in makeready problems and a less positive impression backing. The more common arrangement of a solid pressure cylinder requires an adequate system of external pressure adjustments to allow for variations in stock caliper. Large presses allow for such adjustments by scaled settings on the machine frame to allow for parallel settings to the blanket or for movement out of parallel where job conditions demand. This cylinder contains the grippers which carry the sheet through the printing operation and must therefore be in positive timing with sheet transport and tripping mechanisms. It should not pivot out of engagement because plain sheets must feed through the same line of travel whether tripped out or in.

The inking system contains a large number of rollers to shear the viscous lithographic ink. Most manufacturers incorporate a large drum roller in the system to increase the reservoir capacity, whilst others contend that this reduces the area of surface contact resulting in less shear. Roller coverings must have the surface strength to resist the pull of ink at high speeds and there should be adequate internal cohesion to prevent soft rollers breaking up during the run. As with all machines, ink feed must be continuous enough to avoid sheet-to-sheet variation.

The damping system is comparatively simple in design because water has excellent flow characteristics. The limited capacity of small systems needs fine control over the dispersion of moisture to the printing plate. As with the inking system, all rollers should be adjustable to contacting surfaces and friction driven to the printing plate. A variety of systems are in use and are dealt with in more detail in later chapters.

Automatic tripping is required for pressure lifting; roller lifting; cancelling of feed to inkers and dampers; and to check the feeder. Forme inking rollers must be prevented from leaving solid lines across the plate while the machine is stationary. On the other hand it is not desirable for the dampers to lift automatically as this will cause the printing plate to dry with every interruption to sheet feed and on restarting the dry plate would ink solid. Dampers are therefore lifted manually on all conventional lithographic machines.

Registering time between sheets is an important factor in machine design. Transfer of the sheet from the points of register at the lays must allow sufficient time to clear one sheet before the next is laid. This can be aided by jockey or pull-out wheels which advance the sheet clear of the following one, or guide wheels can swing into position and clear the rear edge of the sheet from the following one. A rotary transfer action is preferred on large modern machines although some still use a swing-arm action which is easier to install but entails a risk to good register at high rates of reciprocation. One company uses a Geneva wheel, carrying three sets of grippers on a rotary cylinder. This

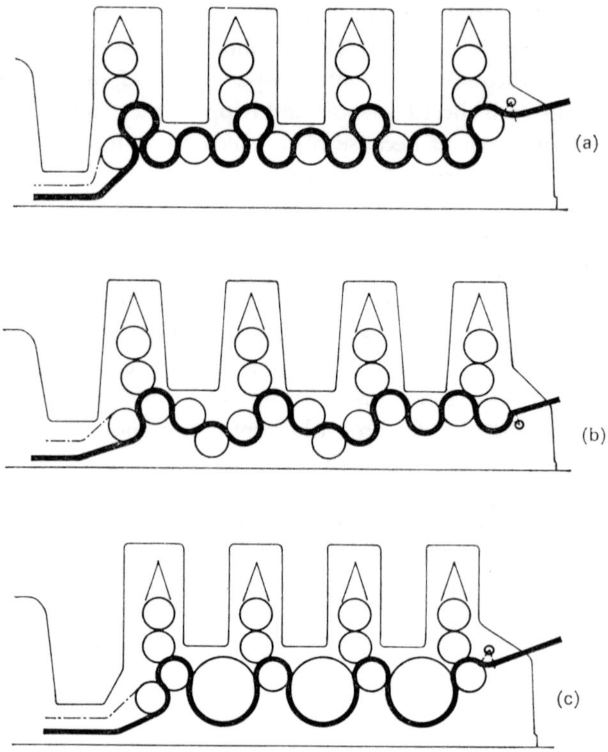

1.11 *Multicolour sheet transport systems: (a) Common level transfer drums showing acute path of sheet travel. (b) Staggered layout of transfer drums gives more obtuse path. (c) Large transfer drums require less space.*

system allows an extended laying time and an open view of the sheets on the feedboard, but the number of alternative gripper changes is extended.

Composite lithographic machines

More flexibility is possible when considering the design of machines with more than one printing unit; the impression cylinder may be individual to one printing unit or it may be common to two or more units; the mode of conveyance between each station can vary with machine design and is important to the type of work to be printed. The more gripper changeovers, the greater the chance of misregister; it therefore follows that the common impression system is the best for register purposes.

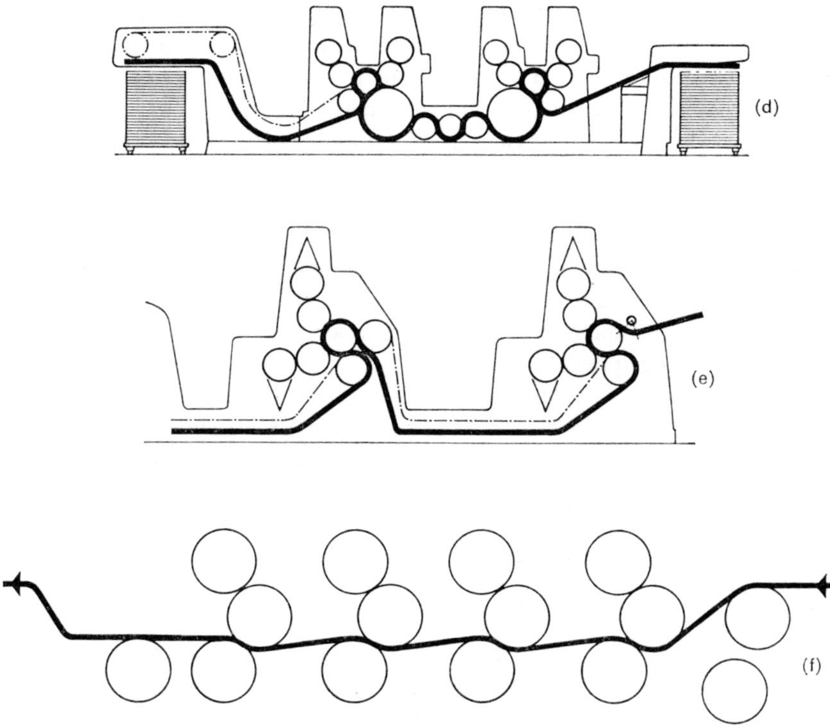

1.11.1 *Multicolour sheet transport systems (continued): (d) Combination system designed for common-impression, 2-colour units, with time for ink to set between units. (e) Another method of achieving setting between 2-colour units using chain trolleys. (f) Pre-register and continuous gripper control to prevent misregister at changeovers.*

Two-colour lithographic presses are often built on the common impression cylinder principle and are based on either a U arrangement or an upright formation. The U-type layout enables both printing units to be fed with ink from above, whereas the vertical plan usually requires one unit to print from below, entailing special consideration in the design of the inking system to obtain equivalent rolling power from an anti-gravity feed. Some two-colour machines utilise a vertical cylinder layout but feed from above on both units. This is achieved by placing the feeder between the printing units and the delivery, and by feeding the sheets face down. Inverted sheet feed introduces more feeder problems but requires less space, which is a particularly useful advantage where a machine is designed for field survey work and requires ease of transportation.

Multicolour lithographic presses may comprise two-colour units linked in tandem for four- or six-colour systems, or alternatively single-colour units can be linked to make odd number combinations possible. Sheet transfer between the units may be by any of the following methods:
1 Three equal-size transfer drums between each unit, with common centre levels.
2 Staggered transfer drums normally with one drum between the first two units and three at subsequent changeovers. The purpose of this system is to avoid severe changes in direction of sheet travel inherent in 1.
3 Large 'jumbo' transfer drums, two or three times the printing cylinder circumference. This method reduces the number of changeovers but requires additional sets of grippers in the drums.
4 Chain gripper trolleys. These are usually employed between pairs of two-colour common impression units to give extended drying time.
5 Endless chain drive passing through all printing units. This system gives continuous gripper control and eliminates changeovers completely.

A further variation is to use double or treble size impression cylinders in single- or two-colour unit assemblies, in which case these cylinders must also carry extra gripper mechanisms.

Inherent problems in the design of multicolour machines are concerned with obtaining accurate register, handling freshly inked sheets without marking and allowing adequate drying time. The method of travel through the machine is therefore important and is dependent upon the type of work to be printed and the substrate. The straighter the sheet travel the better, but space considerations usually prevent this ideal. Board is likely to suffer structurally with bending and, because of flicking through the units, may smear at the print edges. Various devices are used to combat these problems, such as electrostatic systems on the impression cylinders to control sheets by polar attraction, air cushions to prevent sheet marking and the use of ink repellent materials on transfer drums.

Drying times on multicolour machines are rarely adequate and difficult to extend without high cost or excessive use of space. Consequently, this problem is largely tackled by improvements in ink and paper technology, by the delaying influence of anti-setoff sprays and the use of heaters and hot air dryers.

Offset lithographic perfectors are used for bookwork and publications where the work is primarily monochrome text printing. Most manufacturers opt for the blanket-to-blanket principle where both sides are printed simultaneously, the printing blanket for one side being the impression surface for the other. The advantage of this layout is a saving in space and simultaneous registering of both sides of the sheet from a common edge. Some loss of clarity results from the softer impression and the importance of this factor depends upon the type of work.

The blanket-to-blanket system requires one of the blanket cylinders to take over the function of sheet transfer and to incorporate a gripper system. This cylinder should therefore be fixed and not pivot out of impression upon tripping. For example the lower blanket cylinder may be fitted with grippers and the lower plate cylinder made to move out of engagement, while on the upper deck the blanket cylinder will pivot out as normal. The problem in this case is to maintain a true relationship with the inkers and dampers when the lower plate cylinder moves. An alternative design allows the lower blanket to pivot away as normal while the upper blanket cylinder containing the grippers moves towards the lower blanket and rotates so that the change in gripper relationship is kept to a minimum and the plate cylinders need not move. Other perfectors require both blanket cylinders to move, with consequent complications in the transfer gripper arrangements that link this type of press to the rest of the machine.

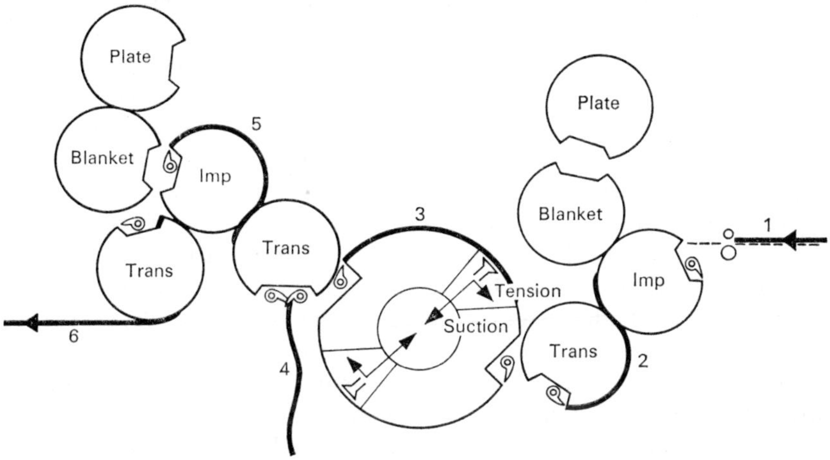

1.12 *Dual purpose 2-colour/perfector, converted for perfecting. 1. Sheet being registered at the feedboard. 2. Sheet already printed on one side is taken by first transfer cylinder. 3. Sheet is held by one set of grippers in the transfer drum. Tail of sheet is held and tensioned by suction. The drum carries the sheet to the second transfer cylinder and continues until the tail of the sheet is taken by one of the sets of grippers in that cylinder. Suction then releases. 4. Sheet is being passed from one set of grippers in this cylinder to the other so that the new lead edge will approach the second unit correctly. 5. Sheet about to be perfected in the second unit. 6. Sheet on its way to delivery.*

Non blanket-to-blanket perfectors require transfer drums between single-colour printing units. As with letterpress perfectors there is a risk of marking the first printed side and it is advisable to incorporate air cushions in the transfer drums or to cover them with ink repellent materials. The principles of true rolling must also be incorporated into the design, as covered in Chapter 4. One machine is designed to be converted from a two-colour machine to a perfector by reversing the sheet travel after the first printing. The patented transfer system between units changes the trailing edge of the sheet to the gripper edge so that the sheet is inverted into the second printer.

Web-fed offset litho machines employ the same basic principles as sheet-fed machines with variations in techniques related to the continuous nature of web feeding and the higher speeds, which may reach 30000 cylinder revolutions per hour. The cylinder gap is negligible compared with sheet-fed printing so that there is no recovery time for the removal of emulsified ink, etc. Ink and water supply must therefore be continuous to avoid variation in print quality.

For perfecting, the blanket-to-blanket system is the easiest to construct and is indeed simpler than on sheet-fed presses where grippers in the blanket cylinders complicate the cylinder lifting mechanisms. Multicolour work may be produced on common-impression cylinder systems, either as satellite or double-cylinder arrangements. To perfect in colour, additional units may be employed or the machine can be designed so that the web can be turned and fed through the same unit side by side with the first printing.

News presses normally have vertical web leads for ease of access and to enable the webs to travel to the folder or other units, while machines designed for colour printing are more likely to have a horizontal lead. Web tension requirements are higher than for letterpress due to the complete contact in the printing nip and consequent distortion of slack paper. Web controls have to allow for this taut web and cylinders accurately packed to give correct web drive. Register between units can be by control of web length between them and by external controls adjusting axial, radial and vertical movements.

Print length is related to the circumferences of the printing cylinders, but variations in print length can be made by interchangeable cylinder assemblies, particularly with blanket-to-blanket assemblies where vertical units can be easily trollied in.

Specialised lithographic machines

The flatbed principle is often employed for proving offset lithographic plates, and such machines can also be used for short runs. An advantage of flatbed printing is that difficult substrates such as tinplate and glass can be handled. These presses vary in sophistication and accuracy from simple, manually operated versions to multicolour machines where pneumatic drive may be used to control the cylinder and gripper operations, with manual or electrical switching of the various control valves.

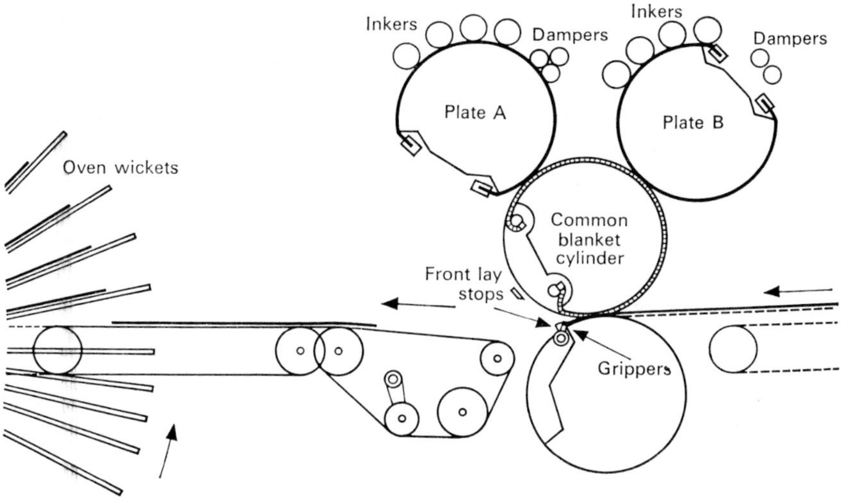

1.13 *Diagram of Y-type 2-colour metal printing machine. Grippers release the sheet so that it will continue in the direction of the travelling oven.*

Some rotary presses are designed to print direct from the plate – mainly for producing posters. These machines have large diameter cylinders and the thicker ink film obtained in printing direct from the plate gives a bolder colour for display work.

Rotary press production of tinplate and other metal sheets is an ever-increasing field demanding specialised techniques and machine design. Machines are designed to give a straight-through travel of the sheets to eliminate gripper control between printing units on multicolour presses. The design must therefore allow for re-registering between units using a tandem layout to print up to four colours. Alternatively a 'Y' layout may be used to print two colours onto a common blanket cylinder which transfers both colours to the metal sheets thus avoiding the need for re-registering. Some of these machines employ a double-size cylinder carrying two blankets in order to obtain better accessibility, although problems can occur if the two blankets vary slightly.

Large transfer drums can be used to handle the metal sheets between printing units and this is of particular advantage where lighter weights of tinplate are required to be printed. Stream feeding is not practical with metal sheets and front lays are contained in the cylinder grippers so that they are travelling in the same direction as the sheets. Sheet drive into the cylinder must therefore be strictly controlled.

On metal printing machines it is also necessary to separate the tripping of plate-to-blanket and of blanket-to-impression, so that each can be operated independently as required. This is done by locating the eccentric bearings on separate cylinders instead of the normal practice of a double eccentric journal on the blanket cylinder. Apart from these special factors the printing unit operates in the same way as machines for printing on paper, but lack of absorbtion by the substrate requires finer tolerances of image control. The printing machine must be linked with a drying oven as an in-line operation. The widespread use of tinplate with a low carbon-steel base makes magnetic control of handling and registration possible.

Engineering problems in offset lithographic press design

Modern lithographic machines need absolute control of cylinder drive and circumference ratios to obtain high quality work. This means that the driving wheels must roll against each other on the pitch circle and the interaction of every tooth must be consistent and free of vibration. Two methods are used to attain this ideal.

The first method requires hardened steel bearers at both ends of each cylinder, machined to the pitch circle and running in contact with adjacent cylinder bearers to provide smooth running. It is claimed that this system is less wearing to the drive gears, which can be of a straight cut type for simplicity, but places strict discipline on impression control. Straight cut gear teeth are in complete contact for only part of their mesh cycle and interruptions to the drive can cause streaks and gear-marking. A better gear design is to cut the teeth so that the tooth flanks follow an involute curve. The contact pattern of two teeth brought into mesh takes the form of a straight line as the gears rotate, and from this contact line the involute can be derived. The angle between this straight contact line and the tangent at the pitch circle is the angle of action. Most manufacturers cut their gears on an involute curve for an action of 20° but others prefer a 15° action angle to keep gear play within closer limits where the distance between axes must be altered, such as when making pressure adjustments to one cylinder only.

The second method, chosen by most manufacturers, is to dispense with the need for bearer contact and to obtain a smooth, accurate performance from the

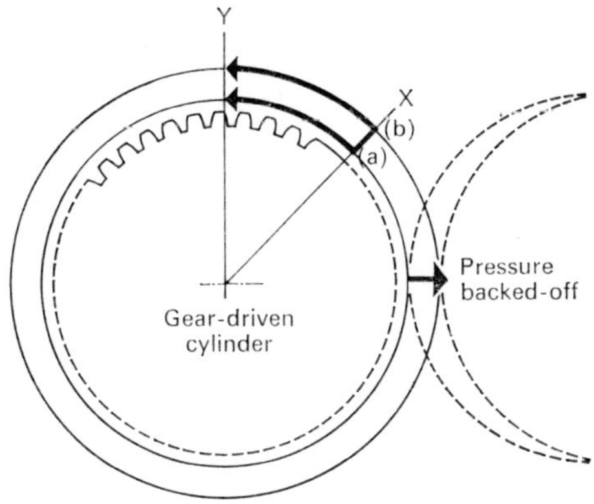

1.14 *Importance of the pitch circle. Overpacking of the plate cylinder at (b) increases peripheral speed of gear-driven cylinder surface in comparison with (a), thus increasing distance of travel between X and Y. This induces friction with the adjacent, normal size cylinder.*

gear drive itself. The gear teeth will be of involute design and cut at an angle to provide helical meshing so that the teeth receive gradual loading and share the drive with other teeth at all times. The greater this angle, the more the loading is shared, but to the detriment of forward drive efficiency, and it

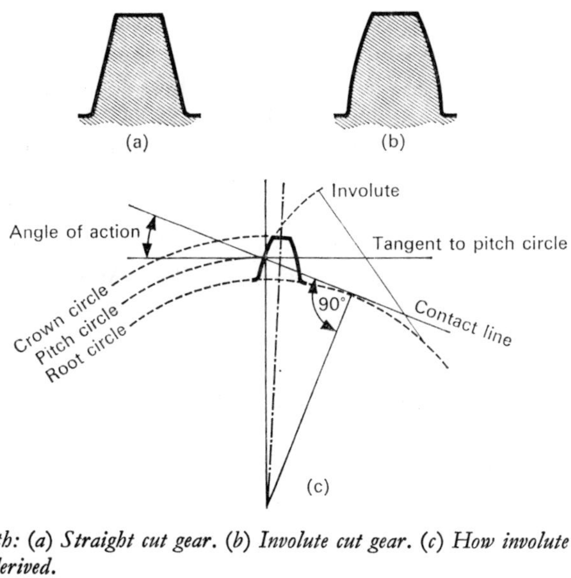

1.15 *Gear teeth: (a) Straight cut gear. (b) Involute cut gear. (c) How involute gear shape is derived.*

follows that the helix needs to be less acute on large presses where power requirements are higher. More elaborate gear designs can include curved shapes and V patterns but the increased efficiency must be measured against cost and replacement.

Another problem affecting mechanical efficiency on large machines arises from the flexing of cylinder shafts under load. This can occur with rollers and cylinders where a load is applied at the ends, causing the centre to bow. Most modern machines have inker and damper roller bearings inside the framework of the machine to minimise shaft length and reduce distortion. Driving gears are placed outside the framework for stability and this also keeps the gears free from dust, lint and ink as they can be safely contained within sideguards.

GRAVURE MACHINE PRINCIPLES

A basic gravure machine will have a printing surface carrier, an inking system with a method of removing surplus ink, a pressure unit, a drying unit and sheet or web feed and delivery systems. Most web-fed machines will also have various in-line units built into the press.

Printing surface carriers are always cylindrical as gravure is exclusively a rotary process working from plates, sleeves or cylinders. The printing surface consists of cells separated by walls, the cells being flooded with liquid ink and the tops of the walls providing a bearing surface for the doctor blade which removes the surplus ink before impression. A continuous surface is provided by complete cylinders which can be lifted into the printing unit and by cylindrical sleeves mounted on mandrels, but where flexible copper plates are used on cylinders with plate clamping and tensioning devices a continuous surface can only be obtained by filling the gap between the plate ends with a segment.

Inking systems are designed to apply the thin, liquid ink to the printing surface so that the minute image cells are completely filled. Application is achieved either by the cylinder revolving, partly immersed in a trough of ink or by an intermediate applicator roller which revolves while partly submerged. An applicator roller is preferred for inking plates or sleeves as partial submersion of the cylinder allows ink to creep under the printing surface. On modern web presses more sophisticated inking systems apply ink to the printing cylinder from a spray rail or by a weir system with a pump circulating the ink to the trough from an off-the-machine reservoir.

After application the surplus ink must be removed from the printing cylinder and this is known as doctoring. The doctor blade is a flexible steel blade, carefully sharpened and polished, usually supported by other blades to give control of its flexibility and clamped in a metal holder. The edge of the doctor blade is lowered on to the printing cylinder, at a predetermined angle, to ride on the cell walls and remove all the surface ink, leaving the cells full. The surplus ink will run back into the trough. The doctor and support blade assembly must incorporate certain adjustments in order to control fully the doctoring operation and to prevent damage to the cylinder; lateral oscillation of the blade should be adjustable and the angle of contact must be variable within certain defined limits.

Pressure units in sheet-fed machines are cylinders designed to carry a permanent resilient packing in the form of a rubber blanket plus a temporary packing which may be paper or other material. The impression cylinder may run in fixed bearings or the bearings may be adjustable so that pressures can be varied. The latter system is essential on machines where printing cylinders of different diameters are accommodated. The gap in the impression cylinder will also house the sheet gripper system, packing clamps and the blanket tensioning device.

On web-fed machines the pressure unit is a rubber covered roller. This may be the only source of pressure or it may be backed up by a robust steel 'bowl' roller or two lighter backing rollers. The pressure unit can be raised or lowered to give correct pressure for printing cylinders of varying diameters and to allow for printing cylinders to be lifted in and out of the press. An alternative system is to have a fixed pressure unit and an adjustable printing cylinder.

Drying units are designed to speed up the evaporation of the volatile solvents which form the basis of gravure inks. On sheet-fed machines pipes with holes or slots in them cross the delivery system and blow hot or cold air to accelerate evaporation. On web machines each printing unit is provided with a drying hood which incorporates air blowers and fume extraction. The web is conveyed within the hood by a number of pipe rollers or by a drum which can be heated by electricity or steam, and there is usually a hot air system with the temperature controlled within close limits. After drying it is important to restore the web temperature to normal as quickly as possible, and this is done by passing the web over chilling rollers through which cold water is pumped.

In-line processes are especially important in gravure press design as all but a small proportion of gravure printing is web fed, and processes normally associated with print finishing are frequently built into the machines. Rewinding, sheeting, folding, cutting and creasing, etc. are considered in Chapters 6 and 7.

Sheet-fed gravure machines

A gravure machine should be robust, vibration free, capable of imparting adequate pressure to ensure good transfer of ink from the image to the substrate and able to maintain good register at maximum production speeds. Sizes of machines vary to cater for sheets ranging from 120 mm × 230 mm to 1 040 mm × 1 400 mm.

Most small sheet-fed gravure presses print from plates of prescribed area and thickness. The plate is fastened accurately around the cylinder by clamping devices at each end which stress the plate around the cylinder by tensioning either the back edge of the plate or both ends. A 360° cylinder is formed by covering the clamping recess with a segment or segments varying in size according to the cylinder dimensions and the area covered by the plate. Print to paper register requires circumferential adjustment on the plate cylinder, but lateral adjustment is not necessary as it is easier to move the paper sideways to the plate. Some manufacturers allow for a card to be positioned between the plate and the cylinder to absorb any dirt particles that might otherwise impress into the plate and cause a printing defect.

The two-revolution principle is used on some smaller sheet-fed machines where the first revolution impresses the sheet and the second delivers it as the impression cylinder rises. Accurate timing is essential between the feeder, gripper mechanisms, impression system and cylinder tripping device on this type of machine. The impression cylinder carries a rubber blanket plus a temporary packing to bring the cylinder to correct printing circumference, and the grippers which control the sheet during the printing and delivery cycles. The gripper mechanisms must allow for varying sizes and calipers of stock. This cylinder is held in eccentric bearings turned by rods on each side of the machine, which can be adjusted so that impression can be altered without changing the packing.

The impression and plate cylinders on some two-revolution machines are the same circumference, which means that the image is inked and doctored twice for each copy printed. On other machines the plate cylinder is twice that of the impression cylinder, having a large non-printing area so that the image is only inked and doctored once for each copy. These smaller machines have a mechanical speed around 3 500 sheets per hour.

Single revolution machines are designed to print from plates or cylinders, depend-

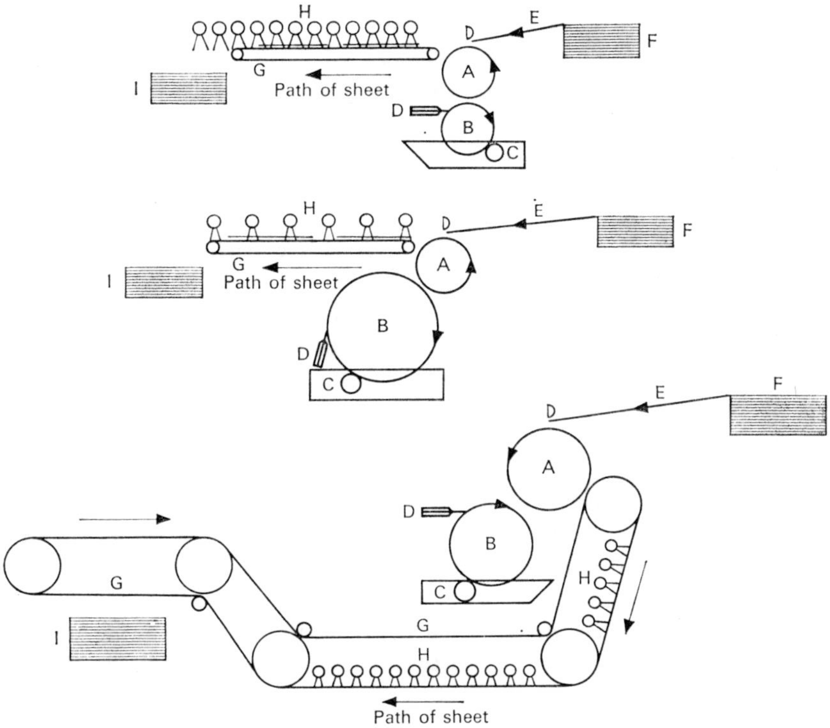

1.16 *Three basic arrangements for sheet-fed gravure printing units: top and centre are two-revolution machines and the lower illustration depicts a single-revolution machine. A, impression cylinder. B, plate cylinder. C, inking unit. D, doctor. E, feedboard. G, tapes. H, drying unit. I, delivery.*

ing on the model. On plate machines the plate and impression cylinders are the same size and revolve once for each printed copy. This means that the cylinder grippers must release the sheet before they reach the feed table in order to take the next sheet. This is made possible by a set of transfer grippers which remove the sheet immediately after the printing cycle.

Sleeve machines are designed to print from a sleeve cylinder, which is a tube of steel covered with a deposit of copper, housed on a mandrel. Copper may be deposited on the base up to the maximum printing circumference, then prepared with an image. On completion of the job the cylinder can be turned and polished to remove the image, and prepared again. This procedure can be repeated many times before further copper deposition is necessary and the technique can also be applied to solid cylinders.

Where cylinder diameters vary it is necessary to alter the packing thickness and the cylinder driving gear, by reference to a chart supplied by the manufacturer, so that the peripheral speeds of the impression and printing cylinders remain identical. It is also necessary to adjust one cylinder in relation to the other and this is usually done by moving the printing cylinder towards or away from the impression cylinder.

Solid cylinder machines are similar in design to sleeve machines but the copper surface is deposited onto a solid cylinder instead of a tube, and the shafting is an integral part of the cylinder. Both types of machine are quite economical of space, with the drying system starting immediately after an impression is taken and carrying on behind the printing unit to continue under the machine until the sheet arrives in the delivery. Machines of this type are designed to run at mechanical speeds of 5 000 to 6 000 sheets per hour and to print on a variety of materials ranging from air mail paper to carton board. Sheet control is similar to that described on letterpress and lithographic sheet-fed rotaries, with swing grippers accelerating the sheets from the feed board to the impression cylinder and positive delivery by means of chain and gripper systems.

Multicolour sheet-fed machines are usually designed with the printing units in line, there being no restriction on the number of units providing there is sufficient floor space. Sheet control between units is by means of transfer drums or continuous chain grippers similar in design to the systems employed in multicolour offset-litho sheet-fed machines. Where space is limited, the machines may be designed with the units arranged vertically or diagonally. The units sit on top of one another in vertical construction, with the normal limit being four units, but with diagonal design the height of the machine can be reduced, with the units arranged so that the sheets descend in a zig-zag cascade.

Perfecting machines can be designed with up to eight printing units coupled together for multicolour work and a sheet turning device between units for printing both sides. Multi-unit sheet-fed machines run at mechanical speeds of up to 6000 sheets per hour with drying taking place between the units by means of cold or hot air blowers.

Reel-fed gravure machines

Reel feeding offers higher production as it dispenses with sheet separation and transport, location of sheets into lays and gripper mechanisms. Web speeds under favourable conditions can be up to 10 metres per second, but in

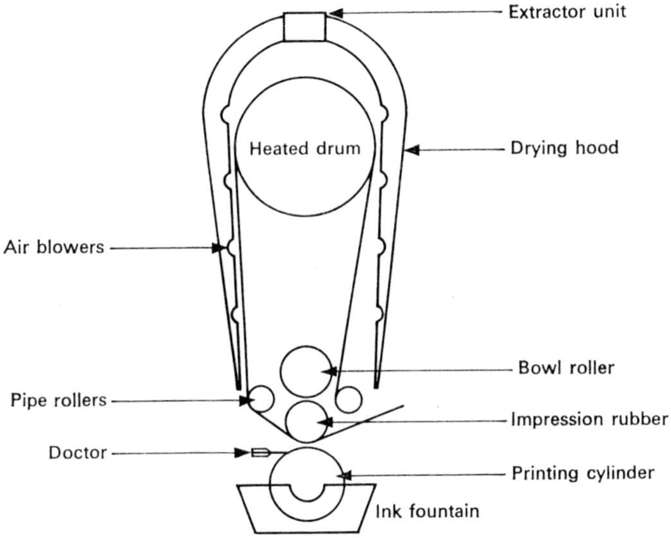

1.17 *Arrangement for web-fed gravure printing unit. The sequence of events in the printing cycle is important, i.e. inking, doctoring, impression and re-inking.*

practice the press running speed may be governed by a number of factors such as the drying time of the ink or the limitations of an in-line operation.

Most web-fed printing units are based on a vertical design with the inking unit at the bottom, the printing cylinder often running in the ink trough, the pressure unit top dead centre of the cylinder and the drying unit immediately above this. The doctor blade must be positioned so that it contacts the cylinder after it has been inked and before the impression nip. Another type of unit construction has the printing cylinder and impression roller horizontally opposed with the drying unit positioned diagonally below them.

Reversible units are made to rotate in either direction so that either side of the substrate may be printed by changing the web lead. With these units it is necessary to house the doctor on either side, according to the direction in which the cylinder is turning. The inking system must be equally efficient in both directions.

The simplest type of press consists of a printing unit with an unwind and rewind device. These units are normally assembled in-line. More elaborate machines will have further printing stations and additional converting units according to the end product required. A common arrangement for packaging machines is to have five printing units which may be used for five-colour work or for four-colour printing plus a lacquer.

Magazine presses in their simplest form may comprise an unwind unit, five printing units and a folder so that a perfected web is produced with four-colour work backed by monochrome. The web is inverted by a 'bay window' in the superstructure, comprising a number of pipe rollers and angle bars designed to turn the web over for perfecting. The size of a magazine machine and the number of units employed will be determined by the number of pages required, their size and whether they are to be printed in colour or monochrome.

Reelstands may be positioned (a) at one end of the press; (b) between printing units; or (c) below the press in a substructure or basement so that the web is fed through slots in the pressroom floor. The first method is used where floor space is limited and no basement exists. The second extends the web lead between units and requires adequate space for a mobile reel cradle running on sunken rails, whereby each reel is located in an identical position by means of a bolt on the cradle which locks into a reinforced hole in the floor. A hoist will be necessary to place heavy reels in the cradle.

Common-impression machines offer the advantages of short web leads (important where costly substrates are printed), good register and economy of floor space compared to separate-unit machines. However, the short web leads cause problems where heat is used to accelerate drying as the impression drum may heat up to the detriment of its rubber covering; also heat transmitted to the printing cylinders may cause ink to dry in the cells before impression. For these reasons the common-impression principle is rarely used for conventional gravure printing, although it is favoured for offset gravure.

Sub-units consists of a printing cylinder with cylinder bearings, an inking system and a doctor blade, which can be removed from the press by sliding onto a trolley. The unit can be taken to an area where the cylinder can be removed, the ink trough cleaned, another cylinder positioned and the unit returned to the press. Spare sub-units can be prepared for the press while it is running. A number of manufacturers offer a basic framework into which gravure or flexographic units can be fitted, depending on the requirements of the job, with drying equipment standard for both methods of printing.

Publication presses consist of assemblies of printing units with one or two folders and a sheeter or a rewind unit. These units are arranged so that a press can produce a multicolour magazine at one pass, or if the full press capacity is not required some units may be used for another job. This is known as 'split-

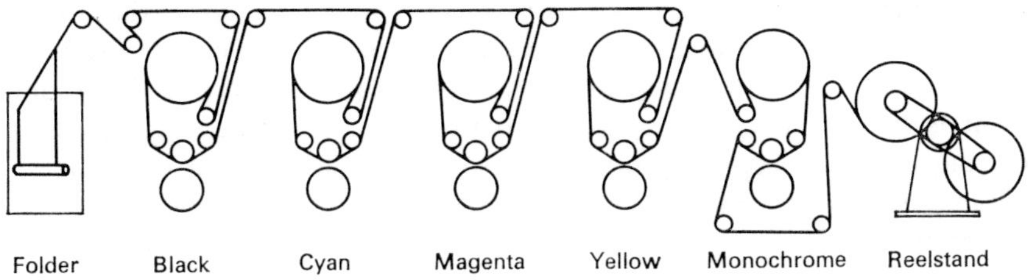

1.18 *Arrangement of units for four-colour gravure printing with monochrome backup. Five separate printing units are required with drying systems on each unit.*

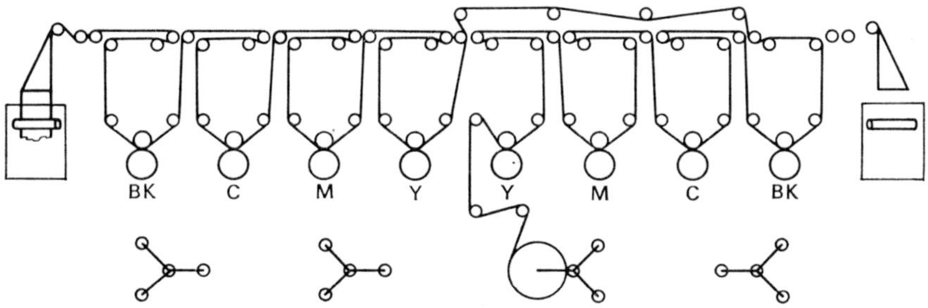

1.19 *Arrangement of eight gravure units, using one folder, printing one reel in four colours backed up with four colours.*

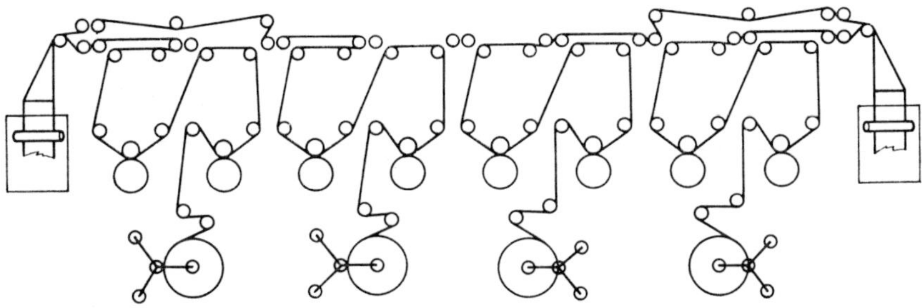

1.20 *Split press principle. Arrangement of the same eight units as in 1.19 to perfect four reels in monochrome, using two folders.*

press' working and requires a separate control system and a second folder. A wide range of permutations is possible where these presses have reversible units.

Carton printing machines are very robust and of unit design. The unwind device caters for automatic changing of large diameter reels by festooning the carton board and allowing the machine to obtain its supply of board from the festoon reservoir while a new reel is spliced in. In-line cutting and creasing units of two kinds are used, the first being a rotary system in which the cutting equipment is housed in cylinders, giving high output but requiring expensive cutting dies. The second is a platen system in which the cutting forme is cheaper but, since the web must be stationary during the cutting process, an intermittent web feed device must be incorporated which reduces the output.

2. Printing surfaces

Images are formed in different ways according to the method of printing – letterpress, lithography or photogravure. Each has its own characteristics, requirements and problems.

LETTERPRESS

A letterpress print is made by inking a relief image and impressing the ink onto paper. The relief is generally produced by removing all non-image areas from a suitable flat surface. The method of removal may be by engraving or etching, and both of these processes may be done manually or mechanically. Once the relief surface has been obtained it may be duplicated by making a matrix from the original.

The technical quality of the printed result is judged by two main factors: the ability of the typecaster or platemaker to reproduce the original copy as a relief image and the ability of the machine printer to ink the plate correctly and transfer the ink to paper. No degree of skill on the part of one can overcome deficiencies in the work of the other.

The letterpress print is characterised by its abrupt and complete delineation between the printing and non-printing areas. There should be nothing but a full and positive film of ink in the image areas and plain paper elsewhere. Intermediate tones between full colour and white paper are achieved as an optical effect by breaking up the image into fine lines or dots of differing surface areas so that the eye confuses ink with paper and recognises the result as a tone between the two. The thickness of the ink film should be the minimum necessary to achieve optimum optical density.

Where a letterpress print falls short of perfection it may be recognised by incorrect inking or imperfect impression. Over-inking or faulty setting of the rollers may result in a building up of ink at the image edges, producing a halo effect at the boundaries. This effect may even be present in prints considered to be of good quality and is sometimes considered to aid legibility by sharpening the outline.

When halo is present, however, it is technically a fault. It may be accentuated by excess pressure which in turn may be recognised by compression or indentation of the paper stock when printing with metal plates, or distortion of the image when printing with compressible materials. Under ideal conditions it should be possible to reproduce detail with fine acutance values and dots or lines with widths to the order of 0·01 mm.

Review of printing surfaces

Text for flatbed letterpress printing may comprise separate type characters cast from individuals matrices and assembled by hand or mechanical means. The width of each character may vary according to the typeface in use but the height from base to printing surface is specified as a constant 23·317 mm. Provided adequate quality control is maintained in the composing room, these three-dimensional characters should be self-aligning when making up lines or pages of type. The non-printing areas between pages are filled with spacing material and the whole locked into a chase by means of expanding quoins.

A forme of type properly composed and carefully aligned should behave as a integral body when correctly locked up. This state will be achieved when the expansion of the quoins within the forme produces forces equal to or greater than the external forces such as gravity. The pressures within the forme should be evenly distributed to avoid undue strain in localised areas, and it is essential that the stresses applied by the quoins should not exceed the elastic limits of the components if permanent deformation is to be avoided.

All other media for printing surfaces will be in the form of plates which may be produced approximately to page size for publications or a correspondingly convenient size for other types of work. A set of plates for any one job must be associated together in a forme, and where the forme also contains type the plates will be mounted up to type height on individual bases. A forme that comprises plates only will usually have them mounted on a common base to avoid the complications inherent in locking up a number of separate blocks.

Original plates may be made from copper, zinc or magnesium by etching, or from plastic materials by photographic exposure and washing out of unexposed areas. Generally speaking, the etching method allows for more detailed control than the washout process. With the washout process meticulous camera work is necessary to ensure first class results.

A third method of producing original plates is by mechanical engraving on plastic or metal with a stylus controlled electronically via a photo-electric cell scanning the copy. These plates may be used directly for printing cheaper types of work or with additional etching, for better quality work.

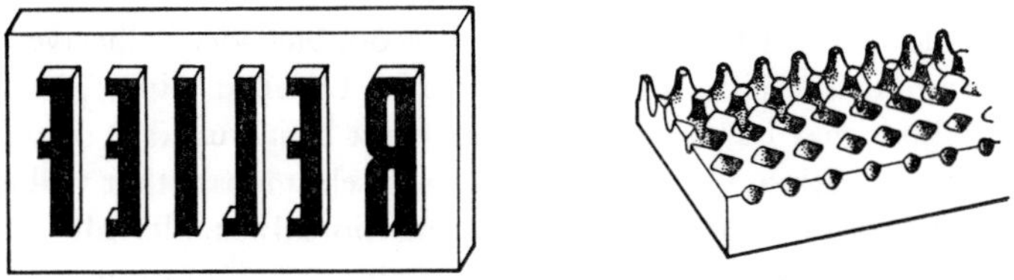

(a) (b)

2.1 *(a) A letterpress plate. The printing image is in reverse and in relief so that only the areas to be impressed receive ink. (b) Portion of a letterpress halftone plate. The illusion of tones (shades between white and black) is created by dots of different areas, which confuse the eye by mixing white paper with black ink.*

Duplicates are made by taking a matrix of the original. The matrix may be used to cast a metal stereotype or to mould in rubber or plastic under heat and pressure. Another method is to grow a copper shell on the matrix by electro-depositing, and to back this electrotype with metal or plastic. Electrotyping produces the most accurate duplication but takes longer and is more expensive than stereotyping. Rubber or plastic plates are quite suitable for the majority of work where duplication is required and there is little demand for metal stereotypes in general printing today.

All the above types of plates may be used for either flatbed or rotary printing, but for rotary machines the plates must be either curved to fit the cylinder or flexible enough to wrap round it. Generally speaking metal plates must be curved while plastic and rubber plates can be wrapped round. This is one reason why there is a higher utilisation of plastic and rubber plates on rotary machines than on flatbeds.

Rotary plates do not have to conform to type height and may therefore be made in much thinner gauges than flat plates. The minimum gauges depend largely on the accuracy with which the relief image can be inked and impressed. A well-designed machine with provision for micrometer adjustment of inking and impression controls should be capable of printing a plate with a relief of 0·1 mm. This requires extremely fine control on the part of the platemaker as well as the printer; in practice therefore plates are etched somewhat deeper. To be stable and durable the relief should not exceed half of the total plate caliper, which means that rotary plates are rarely less than 0·6 mm in caliper and frequently much more. Detailed specifications for printing depth and other relevant factors are given in BS3649.

Plate mounting

Flat plates may be mounted in three basic ways: (i) by nailing to a wooden or partly wooden base; (ii) by glueing with a double-sided adhesive film and (iii) by gripping the sides of the plates with bevelled clips. The first and second methods may be used where plates must be included in mixed formes of type and illustration. Care must be taken to select a base that will not compress under pressure and will remain dimensionally stable. Most all-wood bases will distort in time and it is better practice to mount on metal. Where a double-sided adhesive is used this will give a positive bond to the base but will restrict the machine printer's ability to adjust pressure from beneath the plate.

Where the forme comprises plates only, it is quicker and simpler to lay them on a common base where they can be gripped at the sides by adjustable clips. This system also enables accurate register to be obtained more readily and is particularly suitable where a number of irregular-size plates are used.

Original plates for rotary machines are usually produced as large, one-piece plates, with planning and registering done at the photographic stage. This reduces machine time considerably, leaving the machine printer to concentrate on ink control and impression adjustment.

Duplicate rotary plates may be mounted as curved, rigid segments and gripped at the edges by bevelled clips. For cheaper types of work such as newspaper text where plate register is not critical the plates may be rapidly locked in to fixed positions on the cylinder, but where fine movement of the plates is required for colour register the plates are usually surrounded by adjustable clips.

Individual flexible plates may be mounted on a rotary cylinder by means of a liquid or dry adhesive. This method is suitable for single-colour or non-register work. An alternative is to mount the plates with dry adhesive on to a

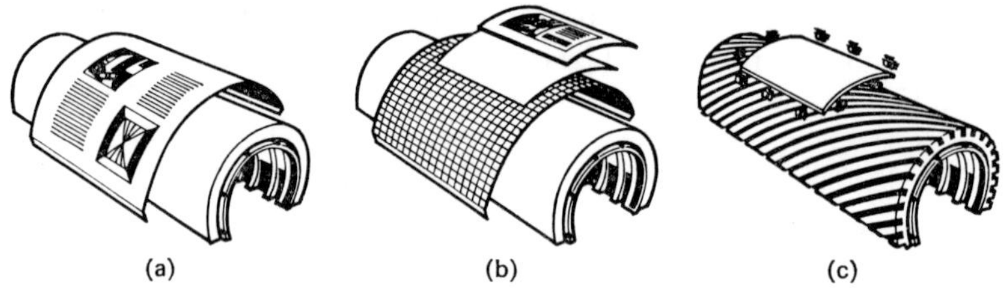

2.2 *Three types of letterpress rotary plates. (a) A one-piece wrapround plate. (b) Plates mounted on to a carrier sheet by means of a dry adhesive film. (c) curved, rigid plates individually mounted on a spiral-grooved cylinder by means of adjustable, bevelled clips (the Warnock system).*

common carrier sheet and wrap the set round the cylinder in the same manner as a one-piece original plate.

Registering letterpress formes

In an efficient establishment any errors in forme register should be detected before machining by proving the forme first in a pre-printing section. Type formes, when correctly composed and locked up are self-aligning, the machine printer's function merely being to position the forme correctly to the paper. Any errors within the forme are normally corrected by the compositor, although the machine printer could be expected to make simple adjustments. If blocks are introduced into the forme, provision should be made to allow for slight adjustments for registering purposes. This can be done by the

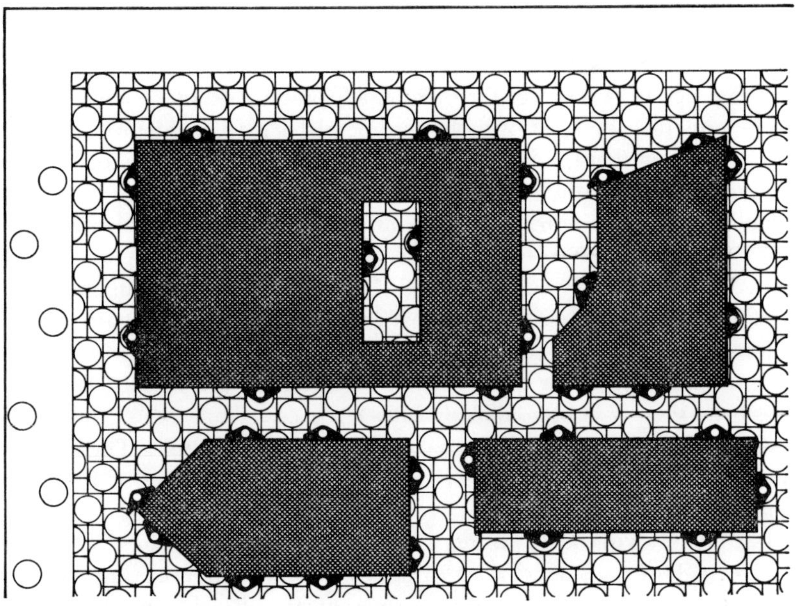

2.3 *Letterpress plates mounted on a honeycomb base. Each plate is surrounded by clips which are finely adjustable by means of a special key.*

inclusion of thinner items of spacing material around them. Adjustment is then made by unlocking the forme and re-arranging the spacing material, taking care to achieve exactly the same tension on the quoins when re-locking. Consistent tension can be obtained by the use of calibrated quoins or by using a torque-controlled quoin key in which the tension applied is pre-determined so that over-tightening is impossible. A better, though more expensive method of positioning blocks is to surround each block with a set of register quoins. These are smaller expanding quoins with precise calibrations and micrometer adjustments so that minute movements may be made to bring the blocks into position. This system is really only warranted for the slight movements required when multi-colour printing from halftone plates.

Where a number of illustrations are contained in one forme, position will be more readily obtained if the forme can be produced as a set of plates to be laid on a common base. A number of patent bases are available on which plates may be held by adjustable clips so that fine movements may be made to any plate without disturbing the rest of the forme. Where a firm undertakes a high proportion of work which needs to be accurately positioned the additional cost of the bases and clips will be more than recovered by time saved in the pre-printing section and machine room.

Colour halftone register

The difference between register of monochrome work and colour halftone must be understood. A set of monochrome plates may be positioned to each other by measurement, within a specified tolerance. Tolerances of less than 0·25 mm are rarely required, and 1 mm is often acceptable.

In colour halftone work three or four plates of different colours are overprinted upon each other, but the dots of which they are composed are not required to be precisely superimposed. In fact the dots are deliberately displaced in relation to each other so that an optical pattern is obtained which will mix the white paper and coloured dots in the required proportions. The degree of displacement is quite small. For instance, in a fairly typical four-colour print there may be 20 dots within a circle of 0·2 mm radius, the average distance between the centres of the dots being about 0·05 mm. Therefore the difference between correct register and misregister is a matter of a few hundredths of a millimetre.

The difficulties in the way of registering colour halftone plates by measurement are: 1, locating a specific set of dots in one plate and relating them to a set of dots in another plate; 2, finding the centres of the dots (they may be of different diameters); 3, calculating their required relationships to each other in terms of distance and angle; and 4, measuring and adjusting their relative positions within the required tolerance. Such an exercise may not be impossible, but it is impracticable, and for this reason colour halftone plates are registered to each other by visual assessment.

This principle applies whether the printing is done on flatbed or on rotary machines. Therefore, a forme or cylinder containing individual plates must have provision for very fine adjustment. This is most commonly managed by surrounding each plate with adjustable clips. The first or 'key' set of plates may be laid down by measurement, within the tolerances specified for monochrome work, and a proof taken to check accuracy. Subsequent sets may be laid in position and compared visually with the key. This visual assessment may be made by overprinting the colours on the key proofs or by the use of registering aids.

Registering aids

The simplest type of registering aid for flat formes is a transparent key produced by printing the key forme on a clear plastic sheet. This key is superimposed on subsequent formes by means of a register bar attached to the machine chase. The plates in the forme may then be adjusted to the key pull by sighting through the plastic. This system avoids the necessity of inking up the forme and taking fresh pulls after each adjustment. For maximum efficiency the plastic sheet must be fairly rigid, stable and free from false movement.

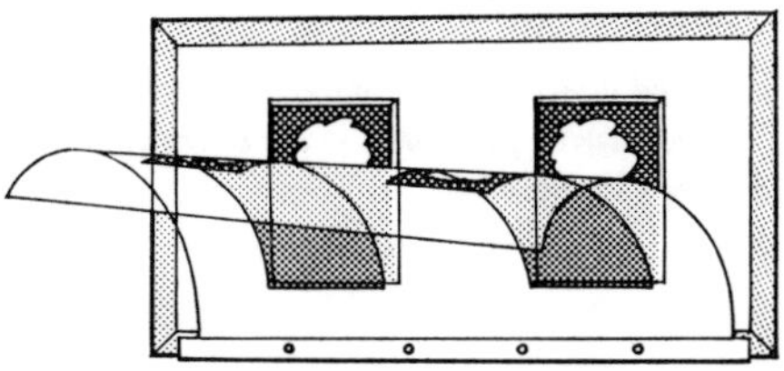

2.4 *A register foil. The transparent plastic sheet is fitted to the chase, the plates inked and the ink transferred to the foil by pressure. The foil can be transferred to another, similar chase to register the next set of plates, or kept until the next working from the same chase.*

A second type of registering aid is the optical device, in which the key forme or print is superimposed by a mirror system onto the subsequent forme. Variations of this principle are used for registering flat and rotary plates. Important considerations with these systems are the avoidance of parallax errors and the faithful superimposition of the optical image in the correct plane.

A special case exists where individual flexible plates are to be mounted onto a carrier sheet to form a composite wrapround set for rotary printing. Again, monochrome plates may be laid down by measurement, the mounting medium being a dry adhesive. This method presents a particular problem for colour registering however, as once the plate is attached to the carrier sheet it is firmly bonded and no movement is possible. For this reason it is essential to establish the precise position of the plate before laying it on the carrier sheet.

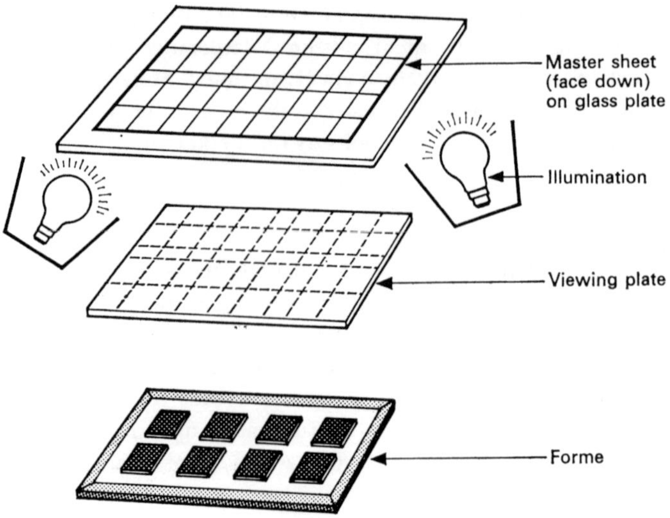

2.5 *Principle of optical registering device for flat plates. The image of the master sheet is reflected on to the viewing plate and may be seen superimposed on the forme by sighting through the viewing plate.*

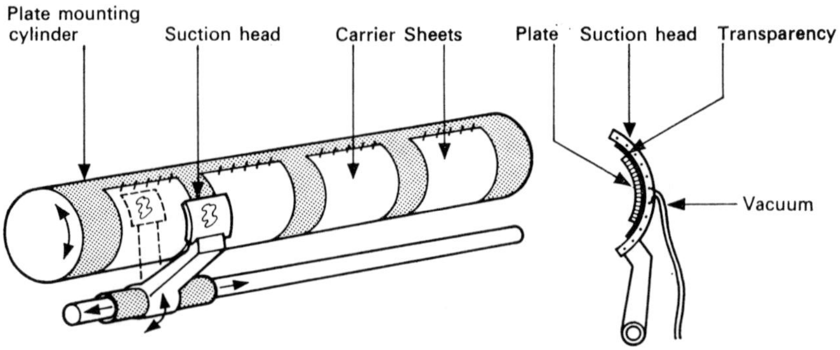

2.6 *Mounting plates in register on carrier sheets; principle of the 'Rota Regista'. For colour halftone work four carrier sheets are mounted in register on the cylinder and a transparency of the plate is taped in position on the Perspex suction head. The first plate is then adhesive backed and registered visually to the transparency, then held to the suction head by vacuum. The arm is moved forward to place the sheet in position on the first carrier sheet. The head is moved along to the corresponding place in line with the next carrier sheet and the procedure repeated with the next plate in the set.*

One way of doing this is to position the face of the plate to a key transparency on a registering device, then lay the plate, still in register with its transparency, onto the carrier sheet. This process is repeated for each colour as shown in fig. 2.6.

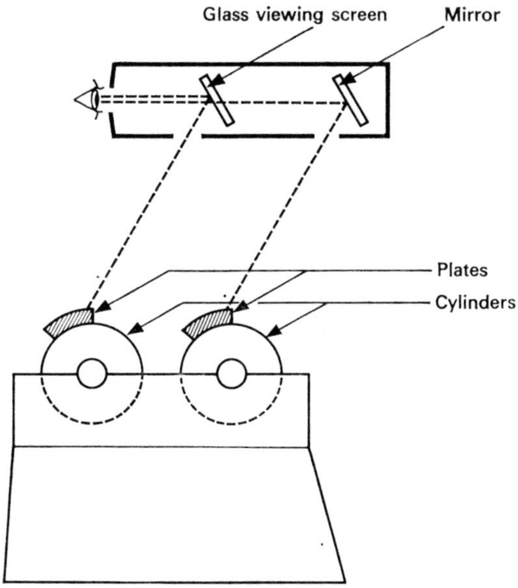

2.7 *Optical registering system for mounting plates on removable cylinders. The plate image reflected by the mirror is seen superimposed on the second image on the glass viewing screen.*

Distortion problems

When a relief plate is wrapped around a cylinder the plate will distort. Distortion is inevitable when a plate is curved because material which does not distort will not bend. If a flat sheet of metal, or a printing plate, is calibrated and curved in a bending device it will be seen that the outer surface stretches while the inner surface remains practically constant. Two significant facts can be observed from this experiment:

1 the smaller the arc to which the plate is curved, the more its surface will stretch, and
2 the thicker the plate, the greater the stretch for a given arc.

It can therefore be seen that the two main factors which influence distortion are plate caliper and cylinder radius.

The amount of stretch which will occur when a plate of a given caliper

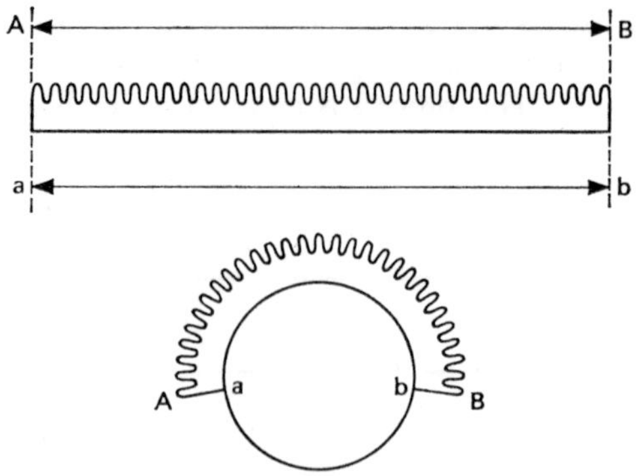

2.8 *The distortion problem. The top diagram shows a flat, relief plate in which the distances A–B and a–b are equal. The lower diagram shows that when the plate is curved around a cylinder, the face and back of the plate are no longer equal in length. Plate thickness and cylinder diameter are two significant factors determining the amount of distortion.*

is curved to fit a cylinder of a specified radius can be expressed as the difference between the arc described by the printing surface of the plate and the arc described by the base, *eg*:

$$\frac{\text{Radius of arc at printing surface (R1)}}{\text{Radius of arc at base of plate (R2)}} = \text{image stretch}$$

For most purposes the radius at the printing surface may be taken as the radius of the cylinder bearers, or on machines without bearers as the pitch circle of the gears. The radius at the base of the plate will of course be the first radius minus the plate thickness.

Example 2 mm wrapround set on sheet-fed rotary with a bearer radius of 190 mm

$$\frac{R1 = 190\,\text{mm}}{R2 = 188\,\text{mm}} = 1\tfrac{1}{94}$$

therefore image stretch = one millimetre in 94.

In practice it is easier to dispense with the whole number and amend the formula to:

$$\frac{\text{Plate caliper (P)}}{\text{Radius at base (R)}} = \frac{2\,\text{mm}}{188\,\text{mm}} = \frac{1}{94}$$

The formula $\frac{P}{R}$ may be used for all plates and combinations where stretch occurs mainly at the face, such as all-metal plates, metal backed laminates and flexible plates on carrier sheets. Where the plate, or at least the base of the plate, is made from a resilient material, part of the plate will undergo some contraction. This means that although the surface of the plate still stretches, it will not stretch to the same extent as a wholly tensile material.

One instance of this is where rubber plates are mounted directly onto the plate cylinder instead of onto a carrier sheet. This will require an adjustment to the formula, which will vary according to the actual material used, but for normal rubbers is usually about 60 per cent.

Example 2·5 mm rubber mounted directly onto a cylinder with a pitch circle of 62·5 mm radius

$$\frac{P}{R} \times \frac{60}{100} = \frac{2{\cdot}5 \times 60}{60 \times 100} = \frac{1}{40}$$

Other exceptions may occur where plates have a fairly rigid printing surface and resilient backs, such as plastic backed electrotypes. With this type of plate, surface stretch can be very small indeed, although it may vary according to the nature of the image. However, it is important to note that when laying down this type of plate on a flat carrier sheet it is still necessary to make the same allowance from the head of one plate to the head of the next as is made for other types of plates. This is because the overall difference between R_1 and R_2 is still the same for any wrapround set mounted on a carrier sheet, and although the stretch may be reduced within individual plates it will be inversely exaggerated in the space between the plates.

Finally it should be pointed out that where original plates are being produced, the distortion problem can be completely eliminated by printing down and processing on the curve. Obviously in this instance it is essential that the curve of the plate during processing should be precisely the same as it will be on the printing press if the image length is to be identical.

LITHOGRAPHY

Senefelder discovered that prints could be taken from flat Kellheim stone by using water soluble gum to prevent greasy ink adhering where it was not wanted. Nevertheless it became common practice to etch the lithographic image slightly in relief so that it was better able to withstand the rigours of direct printing. However, with modern litho plates any etching of the plate surface is primarily a part of the image development process rather than a relief etching of surface metal. For example, with bi-metal plates it is the surface condition of the two exposed metals which decides their selectivity for ink or water, not the contour difference between them.

Louis Poitevin's work on the photo-insolubilising effect of dichromate on colloids produced a printing surface with a continuous tone reproduction of the image which depended upon the degree of water receptivity of a gelatine surface. This process, known as 'collotype', relies on continuous tone photography and is in demand as a means of accurate colour reproduction, although it does not lend itself to high rates of production.

Modern platemaking technology places great emphasis on making the selectivity of the plate surface more precise for finer and sharper printing and more perfect graduation of tone values. To simulate the appearance of continuous tone by conventional lithography the image areas of the planographic surface must be broken up to provide variations in the surface adsorption of moisture. Photomechanical techniques are normally used, in which the image is broken up into dots of varying surface areas by means of glass crossline screens or film contact screens.

Surface qualities required

Lithographic printing surfaces must possess three qualities: sensitivity, selectivity and toughness. Litho stones had a very high degree of sensitivity and selectivity built into them by the lithographer and lack of toughness was countered by the porous nature of limestone which enabled both greasy image and surrounding gum arabic to permeate indefinitely. This ideal lithographic surface has not proved possible in a flexible plate and one or other of these criteria generally suffers.

Sensitivity. The ability of fatty acids to adhere to carbonate of lime is mirrored in platemaking by using the carboxyl group as an adsorbing agent to metal surfaces. These images rely on COOH functional groupings in the fatty acid molecules to bond with the sensitised metal, and maintaining good quality in these plates depends on retaining this bond. Weak images can be strengthened by reinforcing with fatty acids, such as oleic, to improve the carboxyl linkage, but the low polar sensitivity of albumen images makes reinforcement impossible when they become blind.

Deep-etch photographic techniques provide a more sensitive image than surface albumen plates by creating a bond at the metal/image interface due to the etching procedure. A tougher image base, with improved wear properties can be obtained by using lacquers with strong adsorbing qualities.

Selectivity. The lithographic process depends upon the repulsion of grease and water and the printing surface must be made selective to these to obtain image and non-image areas. Monometal surfaces are chosen for their ability

TYPE	DIRECT IMAGE	PHOTO-DIRECT	SURFACE IMAGE Negative ← Working → Positive		DEEP-ETCH	ANODISED	MULTI-METAL Relief ← Image → Recessed	
IMAGE	Fatty acid adsorption and/or absorption	Results from photochemical change of surface	Hardened relief image	Non-decomposed relief image	Image etched into metal	Image surface-integrated	Image metal in relief after etching	Image metal exposed after etching
↓ SENSITIVITY	Very good fatty acids ionic bonding	Good, depending on photochemical material	Usually poor, though some modern presensitized have functional bonding lacquers	Very good, Coating bonds to fatty ink	Very good, Improved with lacquer base	Good, but image can blind if maltreated	Excellent, Copper usually being relief metal	Excellent, Image copper or brass usually
± SELECTIVITY	Fair, depending upon surface metal and treatment	Moderate, Requires careful treatment	Moderate to poor, depending on process	Good, sharp definition, providing development is not hurried	Very Good, initially with retentive gum layer on non-image area	Excellent, Anodic surface is inert	Excellent, Easily revived with chemical treatment	Excellent, Chemically revivable
○ TOUGHNESS	Fair, but fatty image is regenerative except on fine grain aluminium plates	Poor, Short life potential	Poor, except in modern presensitized which are very good	Very good, but image can decompose with poor treatment	Moderate, depending entirely on handling techniques	Very Good, providing anodic surface is not ruptured.	Very Good, bearing in mind that image is in relief	Excellent, providing non-printing area is not ruptured
Ink Image PLATE	Paper, Stone, Zinc, Plastic or Al	Paper, Plastic or Metal	Aluminium or Zinc	Aluminium and anodised	Aluminium or Zinc	Anodised Aluminium	Copper Stainless Steel	Copper or Brass Chromium plus Support

2.9 *Properties of lithographic printing surfaces.*

to be made selective by sensitising and desensitising techniques. The plate is prepared with chemicals to make it sensitive to the receipt of a greasy image, then the non-image surface is desensitised by adsorption of a hydrophyllic colloid such as gum arabic. Adhesion of gum is improved in the presence of an organic acid; *eg* in the case of gum arabic a small addition of phosphoric acid will convert the gum to its free (arabic) acid form with functional COOH linkages bonding tightly to the attracting metal.

Toughness depends on chemical and physical aspects of the lithographic surface; chemical, electrical and atmospheric corrosion of metals can destroy the selectivity of the plate. A simple example is the introduction of too much phosphoric acid in the desensitising etch which will corrode the plate surface and re-sensitise it. Oxidation can occur by reaction with oxygen from the air, or through the presence of more than one metal in the plate surface or the printing unit, creating a 'cell' condition. Physical toughness depends upon the hardness and wear characteristics of the lithographic surfaces and the image itself. This may be entirely structural or may be reinforced by the self-perpetuating mechanism inherent in the migrating, greasy base and correct lithographic techniques.

Lithographic printing surfaces

Direct image plates may be used for work such as large posters where the image is either directly drawn or transferred and relies on a greasy bond to a grained plate of zinc or aluminium. These two metals vary in receptivity, zinc being the better of the two as it has a reactive surface, sensitive to grease and strong image retention characteristics. Aluminium is equally reactive, but provides a less secure base for grease as it is always covered with a thin, tightly-adhering film of aluminium oxide.

Plates for small offset machines may be prepared on paper or plastic surfaces by drawing, by printing directly onto the plate from a letterpress forme or by typewriter. Paper and plastic plates (or masters) usually have a colloid coating with water retaining surface layers which are desensitised by an adhering hydrophillic material such as a phosphoric acid and gum solution. The surface characteristics of absorption and adsorption are fundamental to successful printing; paper plates rely on a degree of absorption with consequent migration of the image over a period of time, while some plastic surfaces have similar characteristics but can be prepared of material which offers a stronger bond to fatty acids.

Electrostatic processes can also be used to provide direct image plates on paper or metal by attracting ink particles electrically and securing them physically through heat fusion. In all cases the resultant image depends on fatty acid linkages with the plate surface.

Direct photographic image plates are still in the development stage. The diffusion transfer system is based on an orthochromatic light-sensitive emulsion transfer paper which is exposed, developed and processed, then transferred to the paper or metal plate by a pressure roll to produce a positive image which can then be fixed. The alternative photo-direct system uses photographic material on the printing plate which is exposed to copy, then developed and processed to produce a plate suitable for short-run work on small offset machines. This method has a promising future because of the reduced exposure time.

Photo-lithographic surface plates make direct use of the coating solution as the image-bearing surface. The surface is coated with a light-sensitive material and exposed through a negative or positive to produce a hardened, insoluble image or a softened, soluble one depending upon the properties of the sensitiser. Press life of the light-sensitive colloid which forms the image base is limited, depending upon its toughness, the tenacity of its bond to the metal and its management on the machine.

The original negative process used dichromated egg albumen as the light-sensitive coating and needed care in preparation to avoid chromic poisoning. Adherence of albumen to metal is purely physical and requires chemical cleanliness at the interface. A grained surface provides a greater area for surface adhesion, but if the depth of grain approaches the coating thickness there is danger of exposing the peaks of the grain after only a little wear. The thickness of the coating can be controlled by varying its viscosity and the speed of whirling, but only within certain limits, as this also alters exposure requirements. Albumen images are not completely insolubilised by exposure and deteriorate under the effect of fountain solutions plus the abrasive action inherent in rolling under pressure. Continuing reaction is also possible, where the image continues to harden under its own inertia after exposure; although where a surface image is produced by hardening of the image through a negative exposure any continued hardening is unlikely to weaken the image, unless by embrittlement.

The modern surface plate is a positive or negative working factory pre-sensitised plate with a long shelf life. Alternatively it may be produced from factory prepared wipe-on coatings applied to fine-grain plates. Such coatings consist of diazo compounds or synthetic resins which, upon exposure, undergo a photochemical change by decomposition or polymer linkage. Positive working systems rely on increasing solubility by exposure. The surface coating retains light sensitivity to a degree and needs to be protected from light at all times by a film of ink. Failure to protect the image in this way may result in its decomposition.

Photo-lithographic reversal processes (deep-etch) take much longer to produce than other systems, but have advantages in production as the coating does not form the image base and press life is consequently increased. Positive or negative working systems are in use and the coatings may be either dichromated gum arabic or polyvinyl alcohol (PVA). The gum based coating does not harden to the same degree as PVA and great care must be exercised when developing with the acid/salt developer to avoid breakdown of the stencil and machine scum at the printing stage. If development is incomplete, a residual film of gum arabic left under the image will weaken it and may not become apparent until the plate is on the press. The PVA system, in common with some pre-sensitised methods, may not give an easy visual assessment of complete development and this can also lead to mistakes.

Deep-etch methods are particularly suitable for use on anodised aluminum plates due to increased resistance to corrosion compared with ordinary grained surfaces plus better surface hardness and improved ability to be wetted by water. The electrically oxidised anodic coating on the plate surface requires a more corrosive etch, hydrofluoric acid, to open the cellular structure of the aluminum oxide layer. Etching does not recess the image as on raw metal, but permits a vinyl base to integrate with the metal surface. The anodic layer is inert to further grease receptance and with correct development will produce good quality printing images.

Photo-lithographic reversal processes (multi-metal) require a photographic stencil to enable selective acid etching through one metal to expose another. The measure of wettability of a metal is the contact angle, determined by a test in which a drop of oil contacts a metal surface under water. Experiments carried out by Adams and Martin provide the following table (p. 53).

Ink receptive areas should show a contact angle approaching zero and non-printing zones should have a high angle approaching 180°. It can be seen that zinc, copper and brass offer better image bearing prospects after physical cleaning and thus can be improved by an acidic preparation. Upon desensitising with gum, aluminium, chromium and stainless steel will show a high contact angle between 170° and 180°. Zinc, copper and brass show a lesser change to between 120° and 130°.

Brass or copper is used as the image bearing surface in multi-metal plates, with chromium or stainless steel for the non-printing surfaces. These metals

Contact angles of image-bearing metals

Metal	*Preparation*	*Contact angle*	*Preparation*	*Contact angle*
Aluminium	With pumice	140	With 1% fluorosilic	50
			With 10% sulphuric	110
Chromium	,,	150	With 10% sulphuric	107
Stainless Steel	,,	110	With 1% nitric	86
Zinc	,,	30	With nitric/alum	30
Copper	,,	77	With 1% nitric	60
Brass	,,	86	With 1% nitric	75

High = water receptive　　　Low = oil receptive

provide satisfactory combinations that can be respectively sensitised with the same chemical treatment. If a bi-metal plate loses its selectivity it can be treated by degreasing, then sensitising the image metal and desensitising the non-image metal. The effects of acids upon metals vary, so that on some combinations one solution is equally effective to both metals as a dual purpose 'reviver'. Common metals retain an oxide surface in air and the frequency of degreasing is related to their chemical receptivity. Thus, zinc and copper are readily reactive, whilst chromium and anodised aluminium possess inert oxide film with natural oil repellency in the presence of fatty acids.

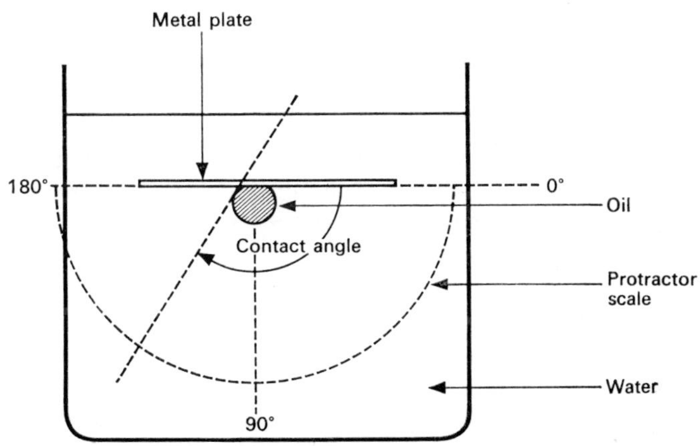

2.10 *Measurement of contact angle. To assess lithographic surfaces a suitable oil is floated into contact under the plate test surface with the test fountain solution. The diagram shows a high contact angle of 125° denoting poor wetting by oil and good wetting by water, ie a hydrophilic surface.*

Damping of the printing surface is the first operation in the printing cycle and will vary according to the selectivity difference between image and non-image surfaces. The better this selectivity the finer control of damping is necessary on the press. Damping unit design is considered in the next chapter, but one other method of moisture application is relevant to image phenomena. Some lithographic proving presses obtain moisture by refrigeration and humidity control. The plate bed temperature is controlled with a refrigeration unit which holds the plate surface at dewpoint so that saturation of air results in deposit of moisture. Low RH would make control difficult because of the low thermostat setting required to reach dewpoint and the possibility of icing up the plate surface, so the humidifier unit is necessary to maintain a reasonable temperature.

Automatic monitoring and control of pH and surface tension of the supply solution during the run will reduce stoppages for plate cleaning. It has been shown by PIRA that lithographic varnishes shed films of oil onto water which they contact, and in the case of the fountain solution/air boundary on the plate surface these films deposit scum on the plate when the machine is stopped. Film formation can be suppressed by reducing the surface tension of the fountain solution with isopropyl alcohol, or more permanently by diol solutions of dihydroxy alcohols.

PHOTOGRAVURE

Photogravure images are recessed into metal or plastic surfaces and the cellular formation is produced by etching, mechanical engraving or moulding. Processing may be carried out on cylinders or on flat plates to be subsequently wrapped round a printing cylinder. Liquid ink applied to the surface fills the cells and the surplus is wiped away by a doctor blade riding on the walls creating the cells, so that the residual ink can be transferred to the substrate under pressure. The characteristics of the printed image will be determined by the image-forming process.

Forms of image

There are three main forms of photogravure image; conventional, inverted halftone and mechanically engraved.

Conventional gravure consists of an image divided into squares of equal size, with walls between them providing the doctor-bearing grid. Variations in tone are obtained by varying the depths of the cells. A print from a conventional gravure image can be seen as ink squares of more or less constant area

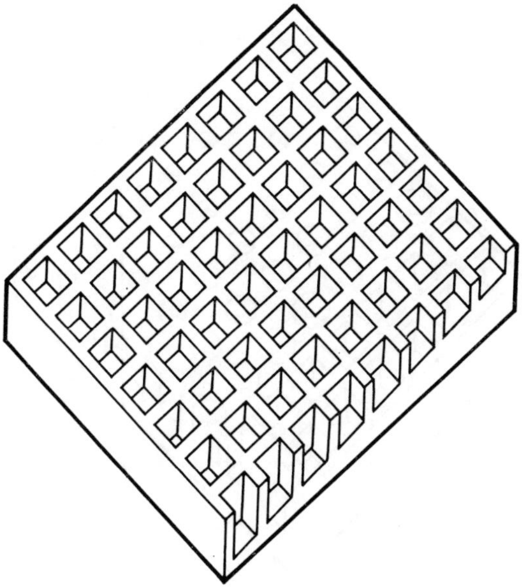

2.11 *Conventional gravure image formation: square cells of equal size but varying in depth. The tonal effects of the print are a result of differing thicknesses of the ink film.*

but variable thickness (constant area, variable depth). In the highlight areas the cells impart a fine layer of ink that is barely perceptible, while the ink in the shadow areas will merge across the grid pattern of the cell walls and increase the density. The highlight cell in conventional gravure is often no more than 2·5 micrometres in depth and on long runs the wear caused by the doctor blade or other factors will result in reduced ink film thickness and a loss of the tonal range, particularly noticeable in the highlights. Cylinders are sometimes chromium plated to obtain longer image life, but this is no guarantee against wear.

Inverted halftone systems have been developed to provide more stable images. By adapting the halftone principle to gravure the cell areas are reduced and their depths increased. The cell walls are, therefore, less vulnerable and, because the cells are deeper, surface wear will not alter the tonal range significantly. There are various kinds of inverted halftone images, but one factor they all have in common is that the cells are not constant in area. Cells may or may not vary in depth according to the process, so that these images can be

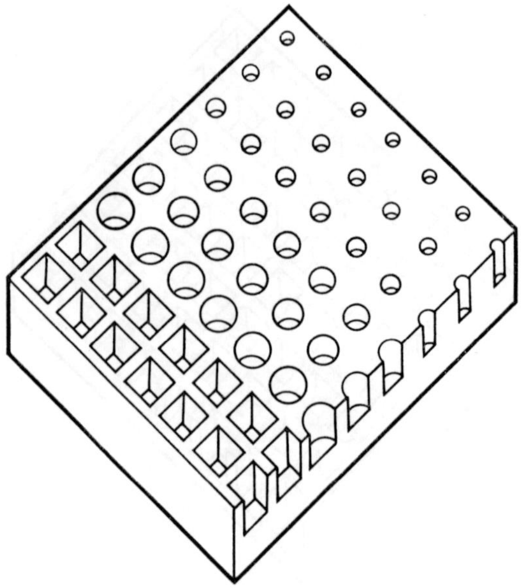

2.12 *Inverted halftone gravure image formation: cells of variable area. Tonal effects are a result of optical mixture of inked areas and white paper.*

broadly classified as 'variable area, constant depth' or 'variable area, variable depth'. The terms 'hard dot' and 'soft dot' respectively, are often used in connection with these images.

Mechanically engraved images have cell characteristics determined by a stylus in an engraving head which is controlled by electrical impulses from a scanning head focused on the copy. The stylus engraves inverted pyramidal cells of varying depths and varying areas and yields soft graduations of tone. Engraving requires a harder copper deposit than etching, which makes this type of image more durable and ideal for long runs.

Screen rulings can vary with all three kinds of image, but the ruling most commonly used in Britain is 60 lines per cm. This gives 3 600 cells per cm^2, and in conventional gravure where the cell to cell-wall ratio is 3:1 or $2\frac{1}{2}$:1, the information available for tone rendering is very considerable. The depth of etch can be varied according to the nature of the substrate, so that a magazine to be printed on newsprint may have a maximum depth of 29–30 micrometres while the same magazine on a less absorbent stock may have a depth of 26–27 micrometres.

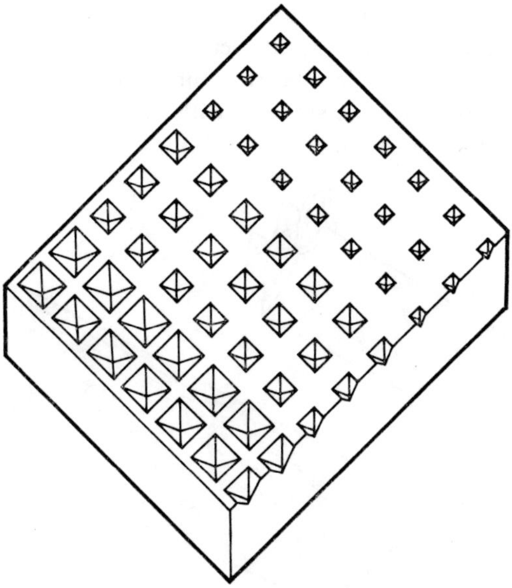

2.13 *Mechanically engraved gravure image formation: cells of varying area and varying depth. The various tones are seen as variations in ink film thickness combined with unprinted areas of white paper.*

Gravure printing surfaces

Gravure plates are made from rolled copper in sizes and calipers to meet the specifications of the machines on which they are to be mounted. For example, a small sheet-fed machine with a maximum sheet size of 400 mm × 520 mm takes a plate 540 mm × 566 mm with a caliper of 0·5 mm. It is important to use plates of the prescribed caliper as variations will result in the plate and impression cylinders having different peripheral speeds, which will cause excessive plate wear and strips of lighter tone across the printed image, known as gear marks. The image surface and the back of the plate must be scrupulously cleaned before mounting on the cylinder as any dirt or protrusion on the back of the plate will cause a pimple to form on the surface, producing the printing fault known as a 'butterfly'.

The ends of the plate must be carefully bent to fit into the clamps on the cylinder. Some press manufacturers supply a plate bending device for this purpose while others build into the machine a special roller which can be set up against the plate after its leading edge has been clamped into the cylinder. As the press is turned slowly the roller forces the plate to the shape of the cylinder and the trailing edge can then be clamped and tensioned. The plate only covers part of the circumference since the plate cylinder must house the clamping system, and the uncovered section must be filled in with a 'gap cover' or 'segment' to provide a bearing surface for the doctor blade. The edges of the gap cover will be smeared with grease to prevent ink seepage when it is bolted into position.

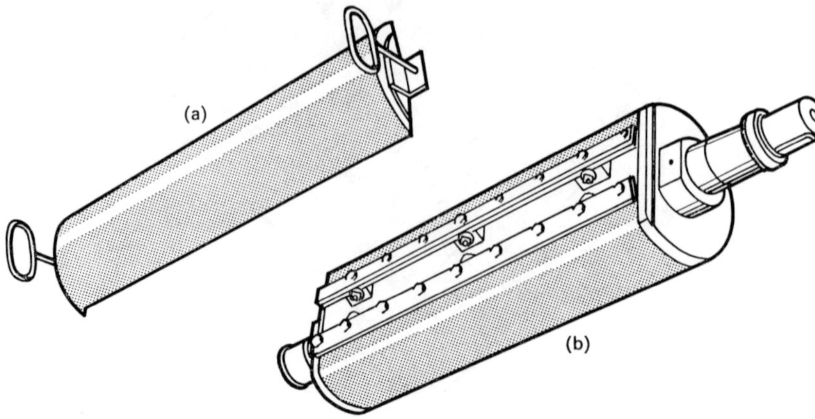

2.14 *Gravure plate cylinder (a) with plate clamped in position and (b) gap cover or segment. A continuous surface is required for gravure printing so that the doctor blade operation is not interrupted.*

Sleeve cylinders are metal tubes housed in the machine on mandrels which can be expanded to hold the sleeves firmly. End collars secure the sleeves laterally and circumferential movement is prevented by a keyway within the sleeve fitting onto a key in the mandrel. On some smaller machines the mandrel and the inside of the sleeve are tapered, with the key and keyway at one end only. Steel is the most commonly used base material for sleeves as it is strong, does not easily distort and accepts copper deposition more readily than other metals. Copper, brass and iron are also used and some light alloy bases are employed when long-distance transport is necessary, although they are more expensive.

Copper is deposited on the sleeve to a diameter slightly larger than the required size. It is then turned and polished in a lathe to obtain the correct diameter and a perfect surface. For sheet-fed presses a thick deposit of copper may be grown to allow a number of images to be etched, run and turned off, before further deposition is necessary. As the diameter decreases the cylinder drive gear and the impression cylinder packing must be changed to keep surface speeds identical. This system is not recommended for multi-unit web-

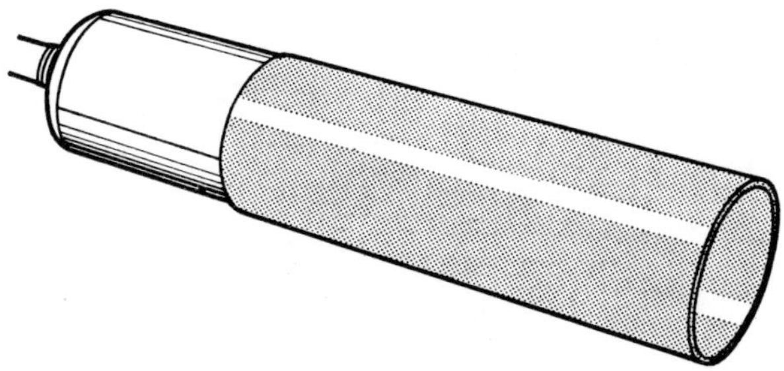

2.15 *Gravure sleeve cylinder. It is only necessary to produce a sleeve or tube with this system, for subsequent mounting on a machine mandrel.*

fed presses because lack of standardisation in sleeve diameters will cause paper control problems between the units.

Solid cylinders are invariably used on web-fed presses and sometimes on sheet-fed machines. The shaft is usually an integral part of the cylinder although some narrow-width presses use removable shafts to reduce capital costs. The cylinders are usually made of steel, onto which a base skin of copper is grown, then turned and polished to a bright finish. A chemical barrier of silver nitrate or silver cyanide is then applied to the copper base and the cylinder returned to the vat for deposition of a further skin which may be of the order of 150 micrometres thick and could take about 1½ hours to grow. A hard, even skin is essential, and this is normally obtained by an agate stone which traverses backward and forward on the cylinder during deposition by a worm thread device, then a final polishing and buffing in preparation for etching. An alternative system uses additives in the copper sulphate solution to obtain the bright copper skin deposit without the aid of an agate stone or a polishing lathe. After the press run has been completed the skin is stripped from the cylinder, leaving a precision base on which a new skin can be grown.

The printing cylinders exercise considerable influence on the web in the machine. If the cylinder preparation is not carried out with a high degree of precision, poor web control and misregister may occur however efficient the web control system may be. When determining the cylinder circumferences on multicolour machines there are three possible approaches; to have the cylinders 1, all the same size; 2, progressively stepped up; or 3, graded from smallest to largest. When the cylinders are deliberately varied in size, this must be the result of a study programme on the press. A progressive difference of 0·5 mm on the circumference may be ideal in one instance, while 1 mm difference may result in web breaks.

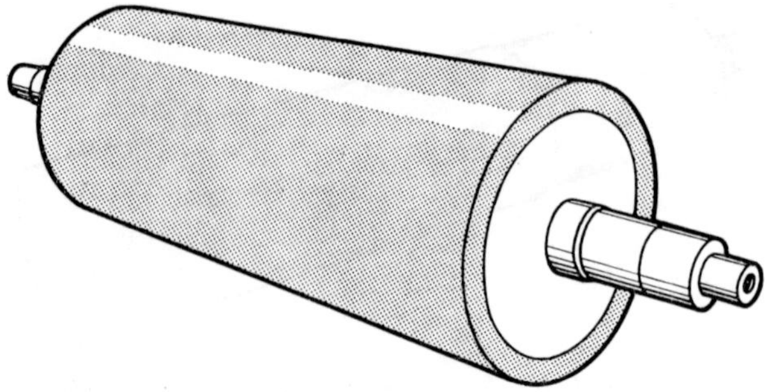

2.16 *A solid gravure cylinder. This is the ideal printing surface for gravure. A continuous image is obtainable, the doctor operation is uninterrupted and ink seepage under a plate or sleeve is not possible.*

Plastic printing surfaces. Colloplas cylinders consist of a thermosetting plastic sheet which is wrapped around a sleeve or solid cylinder and the join sealed. An image is then moulded into the plastic from a relief matrix under controlled temperature and pressure, and the cylinder is ready for printing as soon as the plastic has cooled. A steel-bladed doctor can be used as the plastic is usually resilient enough to allow dirt to pass under the blade without damaging the surface, but if the plastic is scratched it cannot be repaired, although a replacement cylinder can be made very quickly. These cylinders are best suited to inks based on alcohol or water, as inks containing hydrocarbons will soften and distort the plastic.

Uniprint cylinders are similar to Colloplas, but use a special rubber instead of thermosetting plastic. With both types of cylinders pressure settings should be as low as possible consistent with good printing quality, due to the resilience of the image-bearing surface.

3. Inks and inking

A printing ink is a liquid or semi-liquid colouring agent which differs from a dye in that it has a distinct body or viscosity. This enables it to be spread thinly, as a film, on the surface of the printing stock, whereas a dye will impregnate its host substrate by entering wholly into its structure.

INK FORMULATION

Ink formulation can be considered in three parts: 1, the colouring matter, usually pigments; 2, the vehicle or binder to carry the pigments and give the ink 'body'; and 3, modifiers, a whole range of which may be used to either modify the other items or to impart special properties to the ink.

Main constituents of printing inks

Pigmentation is usually provided by insoluble coloured particles of natural or synthetic origin, less than three micrometres in size, finely dispersed in the vehicle. The proportion of pigment in an ink largely determines its colour strength and optical properties, while the efficiency of dispersion affects its flow characteristics and performance on the machine. Dyestuffs are often used as toners in conjunction with pigments, but rarely by themselves due to their transparency, lack of body and tendency to fade and bleed. Exceptions are in gravure and flexographic inks for cheaper grades of packaging where these faults are of less significance, and on metallic foils where transparency can be a positive advantage.

Vehicles are the film-forming agents which bind the colour to the substrate. Letterpress and lithographic vehicles are traditionally made from linseed stand oils, but are being rapidly superseded by synthetic resins which are becoming more readily available, more predictable in quality and in greater variety for special applications. Stand oils are not suitable for gravure inks as the binder must be soluble in a volatile liquid. Resins of natural or synthetic origin are therefore used, ranging from gum rosin and shellac to high polymers. Cheaper gravure inks for printing on papers will probably contain calcium resinate, zinc resinate or a gum binder, whereas inks for packaging applications are more likely to contain resins based on plastic materials such as alkyds, polyamides, phenolics or cellulose derivatives. In all types of inks the vehicle forms the main body of the dried film and controls many of its properties including gloss, toughness, flexibility and protective quality.

Modifiers. Driers in the form of metallic soaps may be added to oil-based inks to accelerate the natural oxidation/polymerisation process. Special resins may be added to supplement vehicle properties such as gloss, adhesion, flexibility and protection. Plasticisers are used to improve flexibility by reducing the rigid cohesion of large polymer molecules. Solvent-type plasticisers are preferred to modified resins, although they are subject to the problem of migration from the ink film after printing. Tinting mediums or extenders are non-pigmented inks which can be used to reduce the colour strength without affecting the structure unduly; gravure printers use 'medium' to control the colour and working properties of ink on the machine, particularly where variations in the etching depth affect the accuracy of the print.

Factors affecting formulation

The formulation of an ink will be determined by reference to four primary considerations: 1, the substrate upon which the ink is to be laid; 2, the machine which will perform the printing function; 3, the printing process; and 4, any special requirements which are desired in the printed end-product.

The substrate. Ink is usually required to remain attached to the substrate and this can be achieved either by partial penetration into absorbent materials or by adhesion. Newsprint and similar absorbent stocks can be printed with a fluid ink which penetrates the paper fibres, while less absorbent papers will require a different bonding mechanism, whereby a sticky varnish forms the body of the ink. The latter will be more expensive than news inks due to the higher costs of the raw materials and the greater expense involved in dispersing the pigments in the vehicle.

Better control of absorption is obtained by coating the surface of paper or board with material having a fine pore structure and capillary attraction for liquids. Inks for these coated surfaces require a solvent phase which disperses into the coating, leaving the more viscous pigment/vehicle phase to form the surface film.

Non-absorbent materials such as plastic films require highly specialised inks to ensure good adhesion. A common approach is to use a liquid ink containing a high proportion of a solvent which will carry the colouring matter and also dissolve selected synthetic resins which can key to the particular plastic. The solvent itself may further assist adhesion by attacking and softening the film surface.

Other considerations of which the inkmaker must be aware are furnish, smoothness, type of coating and surface strength of the substrate.

The machine. Mechanical speed of the machine is an important factor, particularly the surface velocities of the inking rollers and the printing surface. On slow-running machines a viscous, varnish-based ink may be used, while faster machines must have an ink which flows more readily. The slowest machines will use inks with viscosities of several hundred poise, while inks for the fastest machines will be measured in centipoise.

The setting and drying properties of the inks must be adjusted according to the time intervals between printings. On perfectors the reverse side may be printed immediately after the first and setoff will occur if the ink is not suitable. Similarly on multicolour machines where overprinting occurs very rapidly the successive ink films must overprint quickly and 'trap' well, so that each film retains its full colour value. On reel-fed machines where the web is rewound the print must be perfectly dry before rewinding. The ink must also be compatible with the rollers in the inking system and the printing surface in use.

The printing process. Letterpress and lithographic printing inks are similar in that they require a viscous, varnish-like vehicle to carry the pigment to the substrate and to make it adhere. Letterpress inks must be cohesive enough to remain on the surface of the relief images, even on fine lines and dots, without flowing into the non-image areas. The film deposited on the substrate must be full enough to give maximum optical density, and films of less than two micrometres are usually insufficiently opaque. On the other hand, it is uneconomic to run thicker films than are essential for optimum density and too much ink may also slow down drying, affect definition and aggravate printing faults such as setoff and marking-up. For these reasons maximum film thickness in letterpress rarely exceeds 5 micrometres.

The ink film, which is imparted to the printing surface by rollers, is transferred to the substrate under pressure. Not all the ink is transferred, however, as the opposing forces of adhesion to the printing surface and to the substrate immediately after impression cause the film to split, part adhering to the substrate and the remainder left on the printing surface. In offset printing the ink film is split twice – once in the transfer of ink from the plate to the offset blanket and again in the transfer from the blanket to the substrate. Since there is a limit to the thickness of ink which can be transferred under pressure without 'squash' and distortion of the image, it follows that inks for offset lithography must be heavily pigmented to give adequate colouration after a double ink film split.

Another important difference between inks for letterpress and lithography is the presence of water in the latter process, which augments the pigment by becoming dispersed in the ink. Because lithography depends on the antipathy of grease and water it is essential that the ink should tolerate the fountain solution. In consequence the degree of pigment concentration is an important factor in the determination of water tolerance. Without adequate tolerance the image boundaries may expand with furry edges, while too much tolerance will cause contraction and result in a washy print.

Gravure inks contain the same three basic elements as letterpress and lithographic inks, *eg* colouring matter, vehicle/binder and modifiers, but these are further extended by dilution in a free-flowing solvent. This is necessary because the gravure image requires an ink which will: 1, flow readily over the printing surface; 2, be fluid enough to enter and fill the smallest cells; 3, be cohesive enough to remain on the cylinder without flying off at high speeds; 4, have low adhesion to the image metal so that it will wipe easily from the surface; and 5, transfer readily from the cells to the substrate under printing pressure. As well as controlling flow properties, the solvent is instrumental in determining the drying properties of the ink. Other ingredients must be chosen with good solvent release qualities as the diluent has no function in the finished ink film and is therefore required to evaporate completely before the ink is hard dry on the substrate. Incomplete evaporation will result in trapped solvent keeping the ink open and resulting in post-printing troubles such as marking-up, setoff, blocking and bleeding.

Apart from volatile qualities and ability to dissolve other ingredients there are safety factors to be considered when choosing solvents for gravure inks. A concentration of solvent vapours in the atmosphere may be toxic and liable to ignition or explosion under adverse conditions. The flash point of a solvent is the lowest temperature at which a concentration of its vapours will ignite when subjected to a spark or flame, and many solvents are classified as flammable or highly flammable according to their flash points. Their use and storage is closely controlled by local regulations and will affect machine design in such areas as motors, electrical switchgear, lighting and drying systems. Some relevant properties of the more common gravure solvents are tabled in the Appendix.

The inkmaker does not always supply gravure inks ready for printing. It is common practice to prepare concentrated inks with a minimum solvent content so that the printer may exercise control over its working properties. The gravure printer then has two areas of control: 1, choice and quantity of solvent according to ink flow and drying requirements; and 2, choice and quantity of an extender or 'medium' in the form of a colourless resinous ingredient which can be used to reduce the colour strength. The latter requirement may be desirable as a matter of necessity or of policy; *ie* in the event of an etch being unintentionally too deep a proportion of medium can be added to the full strength ink, plus sufficient solvent to maintain correct viscosity, until a correct print is obtained to match the copy. If wear takes place in the image, resulting in shallower cells, some compensation can be made by reducing the proportions of medium added to the full strength ink. Combining these two factors, some firms adopt a policy of deliberately etching deeply to enable them to control printing density by the addition of extenders. It can be argued that this policy requires the use of more ink than is really necessary, resulting in the possibility of drying problems and associated faults. An alternative policy is to etch correctly for inks at normal strength and, in the event of wear, to add extra strong inks with greater pigment concentrations.

It must be understood that the above techniques apply to the correction of colour strength overall. The ideal for which the photogravure printer aims is to have correct a print through the whole tone range, with the lightest highlight cells firmly established on the substrate and the solids printing smoothly without mottling. Local corrections in the event of cells being too shallow, will mean further etching of the areas concerned, while too deep an etch can be locally corrected by applying a lacquer, charcoaling and polishing the image, top etching or employing a proprietary system involving local deposition of metal into the cells.

The end product. If, after printing, the end product is to be folded, waxed, laminated, handled, exposed to strong light or used to enclose foodstuffs, cosmetics, toilet goods, etc. then the ink must be formulated to suit the purpose. Take, for instance, the production of a cardboard carton. The ink must print well on carton board without stiffening the stock, fold and seal without rupturing, not rub off when packed in a box with other cartons for transport, maintain its true colour during the shelf life of the product and be free from odour or any other property which may adversely affect the sale of its contents. Careful discussion with the ink manufacturer is important, and as many details as possible relative to the product should be supplied.

RHEOLOGICAL PROPERTIES AND DRYING MECHANISMS OF INKS

Letterpress

The main properties of an ink which influence its performance on a letterpress machine are those which relate to 1, its viscosity; and 2, its drying mechanisms. The viscosity is a measure of the cohesion of an ink or its resistance to flow. A thin watery ink as used in flexographic printing will flow very easily, while a stiff ink as used for printing non-absorbent paper on a slow-running machine will only flow under stress. A common unit of viscosity is the poise, the value for water at 20°C being one centipoise (0·01 poise) while glycerine at the same temperature has a viscosity of approximately 15 poise.

The stiffest inks may have a viscosity of several hundred poise but it should be noted that viscosity is not a constant factor. For instance, most of the stiffer inks will not flow at all unless a certain minimum stress is applied to shear the ink. In an inking unit having many roller surfaces to mill and shear the ink, the force required to overcome this resistance to flow is defined as the 'yield value' and is measured in decinewtons per square metre. Such an ink is referred to as having an 'apparent' or plastic viscosity, defined as the ratio of shear stress to the rate of flow.

Where an ink has no yield value and the rate of shear is directly proportional to the shearing stress, the ink is said to exhibit Newtonian flow. News ink is one of the simplest of such inks to make comprising about 80% or more of unrefined mineral oil and 10–12% of carbon black pigment, the balance being flow agents and toners. The proportion of mineral oil determines the viscosity, which is an important factor in the performance of the ink. If the viscosity is below 6 poise there is a danger of the ink penetrating right through the paper; above 12 poise the ink will not penetrate sufficiently to adhere firmly to the stock and will rub off easily and transfer to other parts of the machine.

The viscosity of an ink also lowers with increased temperature and with continued shearing such as occurs when the ink passes through a roller train. For these reasons it must not be assumed that a measurement of viscosity in the ink container necessarily gives an indication of the viscosity on the machine. Variations in viscosity due to temperature and continued shearing are reversible so that an ink may be quite stiff in the container or duct, break down rapidly to flow well over the rollers, and having transferred to the stock will again become a cohesive solid. Control of this reversible effect, known as thixotropy, is of great value to the machine printer as it affects ink flow, setting and drying.

Tack refers to the stickiness of an ink. It is measured for printing purposes as the force required to split an ink film between two surfaces. The tack of an ink has a number of implications for the printer: the amount of work done in separating a tacky ink on a complicated roller train may result in a high dissipation of energy and consequential rise of temperature which in turn may modify other properties of the ink as well as the printing rollers and the performance of the machine itself. An ink of low tack may be liable to mottle, may not print sharply and may not adhere satisfactorily to a hard surfaced paper. On the other hand an ink which is too tacky will pull at the surface of a paper, creating the problem of plucking. Suiting the ink tack to paper stock is a basic quality control function. Tack is also a significant factor when overprinting wet inks on a multicolour machine. If the ink on the second printing unit is tackier than the film printed by the first unit then some ink will be transferred from paper to printing surface instead of vice-versa.

The distinction between setting and drying should be clearly understood. An ink is set on the substrate when it has developed a sufficiently firm plastic structure to enable it to be handled gently without smudging. Drying is the complete transformation from the plastic state to solidification, which usually takes measurably longer. The time interval between these phases may be of significance in two ways. Firstly, once an ink has set the work may be further processed without waiting for the full drying period, and secondly, when overprinting, it is usually desirable that the lower ink film should not be completely dry otherwise the superimposed ink will fail to take or may not adhere properly.

The drier modification may consist of two parts: a drying accelerator or catalyst which may be soaps of cobalt, manganese, lead, etc, and possibly a component to retard drying on the press. This second part is usually a wax or solvent anti-oxidant, which will separate from the printed film and not impede drying on the stock. There is an effective limit to metallic drier additions and beyond this proportion the drier merely acts as a reducer.

Modifiers added to improve the printing performance of an ink usually augment the vehicle or liquid ingredient, whereas modifiers added for post-printing requirements often augment the pigment. The particle structure of such materials consequently alters the working properties of the ink. Examples include waxes used to give scuff resistance to the printed film, and starches introduced to prevent setoff. Where it is necessary to employ additives, it is best to leave them until the last pass through the printing machine so that overprinting will not be affected.

Inks based on mineral oils, which do not readily oxidise, do not dry in the generally understood sense. The thin, liquid part of these inks will enter an absorbent material very rapidly due to the capillary action of the fibres. The more solid parts of the ink remain on the surface of the stock, and there is no chemical change. This is why the print from newspapers will rub off.

Rapid-setting inks employ a combination of vehicles. The thinner part is a solvent for the resin and effects a fast penetration into the printing stock, leaving the pigment and resin as a thin film on the surface. Setting by this method takes place in a matter of seconds rather than hours and may be accelerated by partial evaporation of the solvent. Even faster setting in letterpress printing may be obtained by using a combined vehicle in which the liquid part is precipitated by the presence of moisture which is usually present either in the printing stock or the atmosphere. The vehicle for these inks is based on synthetic resins of comparatively high acidity dissolved in a glycol. The addition of a small, but critical, amount of water will cause the resins to be precipitated from the solvent which will rapidly enter the substrate, leaving the resin to bind the pigment on the surface.

For high-speed printing on letterpress web machines, the drying process may be accelerated by the inclusion of a volatile solvent in the ink. Solvents dissolve quick-drying resins in the ink which set almost immediately once the solvent has evaporated. The printer can exercise control over the evaporating process by removing solvent saturated air and by temperature adjustment. (Chapter 6 for details of drying mechanisms.)

Lithography

As with letterpress inks, lithographic printing requires sticky adhesive characteristics in the vehicle which transports pigment to the substrate. Coarse pigment particles will cause rapid wear of machine plates, and most natural inorganic pigments are not suitable. Because of the presence of water in the printing process, pigments must be resistant to fatty acids and acid salts present in the fountain solution and have no detrimental chemical action on the plate. They must also have a greater affinity for being wetted by a greasy varnish than by the damping solution. The varnish itself must tolerate the presence of water, or emulsification will cause a breakdown of the ink structure and de-wetting of the pigment. Tinting will then occur due to the pigment being washed out of the vehicle, and any water solubility of pigment media will cause it to bleed into the fountain solution.

Successful transfer of ink involves preferential film splitting in a system containing water. Good water tolerance requires a high level of internal cohesion in lithographic inks. This ensures a good hold-out or 'finish' on paper stock and reduces the amount of absorption of the ink film. It is necessary to introduce a liquid part to allow firstly for filtration into the paper surface and secondly to enable a reduction in ink film consistency for better ink transfer. It can be seen, therefore, that all litho inks are non-Newtonian in behaviour. The need to obtain good flow properties at the printing surface are of paramount importance and litho machines are designed with an extensive roller train to shear the ink adequately. Viscosity measurements are assessed once the yield value has been overcome, and defined as a measurement of plastic viscosity.

It is an advantage to emphasise the non-Newtonian behaviour in some systems by imparting a false or thixotropic body. The semi-rigid structure will break down on agitation so reducing the effective plastic viscosity at high rates of shear without a parallel reduction in tack. The advantages of such systems is the ability to reduce viscosity on the press without a high level of solvent addition. An added advantage of the high apparent consistency is improved water-resistance and consequent limiting of the film of oil on water normally occurring with lithographic printing.

Lithographic inks suffer a variety of inhibiting effects to their drying performance not met by polymerising letterpress inks. Water tolerance is necessary for printing but will slow the drying rate, particularly when excess moisture releases inhibitors from the paper stock. At an acidity of pH 5 or less, metallic driers are absorbed from the ink by the moist, acid paper. Fatty acids from the ink and acid salts from the fountain damping solution augment this inhibiting effect. Pigment is more vulnerable in litho inks, which need a higher pigment concentration to counter loss of filmweight in double transfer via the intermediate blanket. Pigments must be fine and thoroughly dispersed to ensure good wetting by this vehicle and uneven drying must be expected if particles are allowed to agglomerate. Multi-colour wet-on-wet printing produces a sandwich of such variables, including moisture, which must be dispersed. Choice of ingredients is governed by the requirements of the printing plate and fountain solutions, the need for high pigment content, and compatability with rubber blankets and rollers.

Gravure

The flow properties and drying abilities of gravure inks are largely affected by the qualities and proportions of the solvent content. An appreciation of the functions of the solvent in a liquid ink is essential if the nature of these inks is to be understood. The solvent initially controls viscosity, which is the prime factor governing the ability of an ink to flow freely into the cells of the printing image and to be removed from the surface by the doctor. It follows that the viscosities of gravure inks are measurably lower than those for letterpress and lithography. A stiff gravure ink, for slow-running sheet-fed machines, may have a viscosity of about 25 centipoise, while an ink for high-speed reel-fed machines may be as thin as 12 centipoise.

The viscosity is affected by temperature and continued shearing as are letterpress and lithographic inks, but a third significant cause of viscosity variation in gravure inks is solvent evaporation. In an open inking system there will be a continual loss of solvent into the atmosphere and this solvent must be continuously replaced to maintain correct working properties. For this reason an enclosed system is preferred so that solvent loss can be minimised and added benefits built in by using a circulating system which will maintain the ink level from a bulk supply as well as filtering the ink and monitoring viscosity.

Shearing of the ink from the image surface should occur as late as possible in the printing cycle to minimise solvent loss from the ink in the cells, thus keeping it as wet as possible prior to transfer to the substrate. After transfer of ink from the image to the substrate in the impression nip there may be some ink left in the image and, due to the volatile nature of gravure inks, there is a possibility of this residual ink drying in the cells which will affect the printing quality of subsequent impressions. For this reason the application of ink should occur as soon as possible after the impression nip so that residual ink is quickly re-wetted. Where etch drying does occur, the machine printer may try to correct the fault by the addition of extra solvent. This remedy usually reduces the print quality by flattening out the tonal range and, in extreme cases, causes mottle. A better approach is to add a small quantity of a slower-drying solvent to keep the ink open longer.

When diluting ink it must be understood that not all the ingredients are equally readily wetted by the solvent. Precipitation may occur with shock dilution of an incompatible diluent and some inks may require stirring for several minutes before the constituents are fully integrated. These problems are particularly applicable to co-solvent inks which are based on two different types of solvent such as a hydrocarbon and an alcohol. This type of ink enables the manufacturer to dissolve certain synthetic resins in an appropriate solvent, then to obtain a suitable viscosity by further dilution in a second solvent. The proportions of the solvents must be kept within specific limits to avoid precipitation or coagulation.

An empirical method of establishing and monitoring ink viscosity on the machine is by means of a flow cup. This is an open container with a capacity of several millilitres and having a lower orifice of a few millimetres. The cup is filled with ink and the time taken for it to flow through the lower opening is noted. This will give comparative flow rates which can be roughly correlated to viscosity values. If an ink flows too quickly through the cup (say less than ten seconds) differences in time values will be difficult to record and adjust, while too slow a flow rate (say over one minute) may allow evaporation to take place and ink to dry in the cup, as well as protracting the operation. For these reasons a range of cups should be available and one selected which will give readings of about 20 seconds for correct viscosity inks. Manufacturers ascribe arbitrary numbers to the different cups, usually with the lower numbers indicating the smallest orifices. For instance, a number 1 cup would be used on the fastest-running machines, requiring inks of low viscosity, while sheet-fed machines which use a more viscous ink would be better served with, say, a number 3 or 4 cup.

Once an ink has been approved for a job, a chart should be kept giving details such as the ink makers name, colour and reference number of the ink, quantity issued each time, the proportion of medium (if added), the viscosity and the type of flow cup used. The chart should be a complete reference for the ink in respect of the machine room application.

Inks for special purposes

The influence of the end product has already been mentioned. This will determine whether special properties are required in the ink, such as high or low temperature tolerance, flexibility, gloss or matt finish etc, and may also specify particular freedom from possible faults such as odour, fading, chalking, bloom and rubbing off. In addition to these considerations, manufacturers may also be asked to supply inks for special effects or particular purposes and some of the more common of these are described below.

Metallic inks. These are based on finely powdered metals, usually aluminium or copper/zinc alloys, in a suitable medium. The particles are in fact small flakes of metal about 2 micrometres in caliper and with areas of about 200–300 square micrometres. The success of a metallic ink lies largely in its ability to orient these small flakes as the ink dries, so that maximum light is reflected from the flat surfaces. When working with these inks it is advisable to use minimum quantities at a time and to replenish frequently to avoid atmospheric contamination. The minutest traces of sulphur in the atmosphere or the substrate may cause darkening of the ink as metallic sulphides are formed.

Fluorescent inks. As with metallic inks, the secret lies mainly in the pigment. Fluorescent materials possess the property of photoluminescence, which means that they have the ability to reflect more light of a certain wavelength than actually falls on the print. A limited range of colours is available and thick ink films are necessary to obtain the best results. Maximum fluorescence will be obtained when printing on a white base as the pigments are often transparent, and it may even be necessary to print a double layer for a really striking effect.

Doubletone inks are used to produce a softened and graduated effect with monochrome halftone prints. They are based on a vehicle containing an oil-soluble dye in combination with a stable body-pigment of a different colour. Printing is carried out in the normal way, but over a period of several hours the dye migrates from the medium due to the capillary forces of the printing stock, to form a softening halo around the dots.

Magnetic inks contain a high loading of iron oxide pigment which enables them to be magnetised when exposed to an energising field. They are used in automatic data processing or sorting systems where symbols printed with magnetic ink can be quickly magnetised and sensed by a reading head. Printing with this type of ink requires extremely accurate control of ink film thickness to ensure adequate signal strength and completely faithful reproduction of the image to ensure accurate recognition of the characters by the data processing equipment. The slightest tendency of the ink to setoff or squash will affect the interpretation by the reading head.

Security inks. For cheques, bonds, stamps and other securities, inks are required to be particularly reliable and permanent as well and to withstand any attempts to alter or deface the document once it is in circulation. A common precaution is to incorporate a fugitive dye in the ink which prints the background tint so that any attempt at erasure or alteration will cause the fugitive dye to spread and discolour the surrounding area. These inks often bring problems when used on conventional printing machines as the vehicle for the fugitive dye is usually a volatile solvent which may cause drying problems, and must be checked for compatibility with inking rollers and printing plates.

Collotype inks need soft pigments which will not injure the planographic, gelatine surface. They should be stiff, well pigmented and sufficiently hydrophobic to resist the water and glycerin used in the process. High cohesion is essential for good tonal reproduction, and this applies equally to modern systems of continuous-tone lithography.

Moisture-set inks have a finely balanced formula in which precipitation of the binder occurs in the presence of moisture. The vehicle is a water-insoluble resin dissolved in a water-miscible solvent such as a glycol. These inks are only suitable for porous stocks as the setting mechanism requires the diluted solvent to drain into the substrate. Moisture for precipitation of the binder may be introduced by saturating the post-printing atmosphere with spray or steam, but for well-balanced inks the moisture normally present in hygroscopic stocks or the atmosphere can be sufficient to ensure precipitation. The main advantages of this system are rapid setting and drying together with freedom from odour, strike-through and setoff.

Letterpress offset inks need not vary greatly in characteristics from lithographic inks, although low tack is essential as there is no moisture present to lubricate the blanket for strip-off. Moisture-set inks are used for printing heavy lined boards for cartons where the liner tends to be oleophobic because of the aqueous adhesive used. Being equally hydrophyllic, the surface is attractive to moisture-set inks which cannot be printed by lithography.

Metal decorating inks rely on the use of oven equipment (described in chapter 7) to accelerate the process of oxidation/polymerisation and may effect a curing of thermosetting resins at a predetermined temperature. Pigments must therefore be resistant to fade in the oven in addition to normal lightfast requirements and inks used to decorate pressurised containers for foodstuffs must also resist fading during the sterilising and pasteurising processes. Resistance to fading through light, heat and processing are not necessarily correlated and must be separately defined. Solvents are only added to produce good flow and transfer and should be dispersed by evaporation in the oven after printing. Adhesion to the substrate is more important than tensile strength of the ink film and there should be sufficient distensibility to take up strains resulting from fabrication of the metal.

Relief plates are used for offset printing to metal where the design of the work makes this a practical proposition. Problems of overprinting wet-on-wet ink films are thus reduced by the elimination of the water variable and its deteriorative effect on ink transfer to the non-absorbent substrate. Solventless ink systems are an advantage and inks which solidify on exposure to ultra-violet sources bring obvious benefits in a printing process which does not rely on solvent adhesion. Ultra-violet setting between printing units will also prevent second-impression build-up on the following unit and the absence of solvent will speed stoving schedules as well as lessening the possibility of ink drying on the press.

Inks for metal foils. Most foils are made from rolled aluminium and acquire greasy surfaces during manufacture. The surfaces are cleaned by washing in a degreasing solvent, then lacquered with shellac, vinyl or cellulose nitrate to make a suitable base for printing, with good adhesion properties. For gravure and flexographic printing the foil should be softened by annealing, and solvents and resins should be chosen which are compatible with the particular coating in use. Lithographic inks for foils must also have an affinity for the coatings and are usually linseed oil types. In this instance solvents should be almost non-existent as absorption is at a minimum. Prevention of setoff requires high pigmentation and minimum press moisture. If foil is laminated to board, it acts as a barrier to moisture evaporation and oxygen circulation; the ink film must therefore have a tenacity for oxygen retention to aid stack drying.

Inks for cellulose films. Films made from regenerated cellulose have different surface properties from papers, particularly the lack of capillary structure. A plain, transparent cellulose film (PT) will be comparatively absorbent and can be easily wetted by most gravure and flexographic inks. Printing procedures will not be very different from those for paper stocks, although the reduced absorbency may slow down production and require a more efficient drying system.

Other cellulose films have coatings of moistureproof lacquers based on vinyls, synthetic resins, cellulose nitrate or waxes, to improve their packaging properties. For these, inks should be formulated with solvents which will readily wet the surface and binders which will key to the coating due to similarities in molecular structure. Temperature control is also important, as heat will help to soften coating materials, aid adhesion and seal the ink vehicle to the substrate as well as evaporating the solvent after printing. Coatings containing waxes may be softened by the inclusion of small quantities of ketones or proprietary surfactants to improve their wetting properties. Failure to achieve thorough wetting of the substrate will result in incomplete coverage and the printing faults of pinholing, beading and crawling.

Inks for plastic films. Most of the materials used as plastic films are high polymers of low density, good transparency, excellent dielectric properties and chemically inert surfaces. Surface inertia is the main problem for printing and may be overcome by flame treatment or chemical means. However, the most common method is the corona discharge system in which the film is passed round a metal roller covered with a dielectric layer and a set of electrodes positioned nearby, so that the air above the film is made conductive when a high voltage is applied, causing sparks to jump the gap to the film surface. As with the coated types of cellulose films it is essential to choose ink solvents which will readily wet the surface and resins of similar molecular structure to the substrate. Temperature control is more critical as some films, such as polyethylene, have low melting points. Other faults which may occur are migration of some ink constituents through the plastic film and peeling of the printed image due to poor adhesion.

DESIGN OF INKING SYSTEMS

Letterpress and lithography

The inking systems on the majority of letterpress and offset-litho machines can be divided into four sections: 1, the container; 2, the metering system; 3, the distribution system; and 4, the printing surface inking system.

The container. For most sheet-fed machines the duct system is used. This comprises a steel roller which turns against an adjustable steel blade; the gap between the blade and the roller determines the thickness of ink film metered out. The blade is usually situated below the roller, but where a large capacity container is required or other design considerations demand, it may be above the roller.

On high-speed news and magazine machines the free-flowing ink may be contained in an enclosed box from whence it is pumped through small pipes

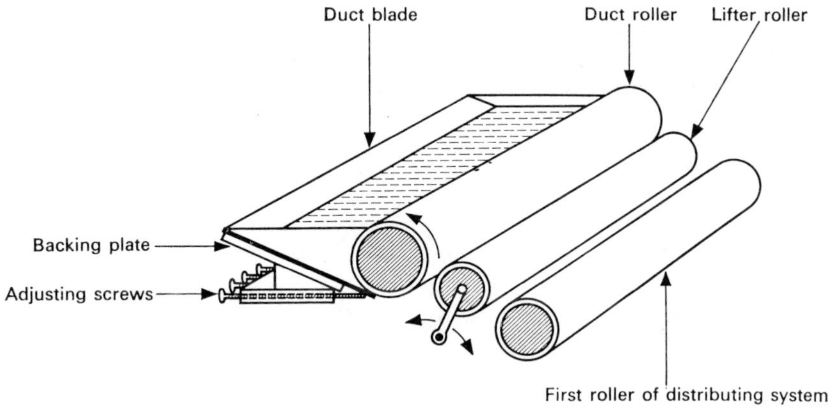

3.1 *Conventional duct. The ink is contained between the duct blade and the duct roller. Localised adjustment to the ink film metered out is by adjustment of the screws below the blade. Overall adjustment of ink flow may be made either by adjusting the rate of rotation of the duct roller or by varying the length of time which the lifting roller dwells against the duct roller.*

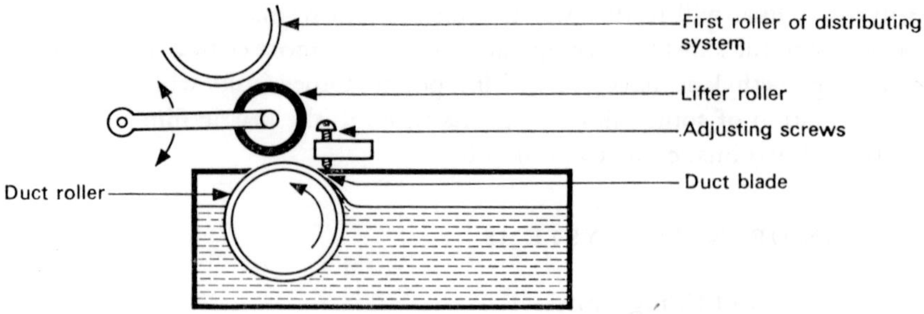

3.2 *Overshot duct. This system is used to meter fluid inks on fast-running machines such as news and magazine rotaries. The container can be fully enclosed and of large capacity.*

to the distributing system. Where completely fluid inks are used, as in flexography, the container may be a simple trough in which the first roller of the distribution system runs. Stiff lithographic ink requires motorised agitators to stir the ink in the duct, so preventing it from backing away from the roller.

The metering system. This may be obtained either by overall control of the ink film which is fed out of the container, or by such control together with lateral adjustment of localised strips of ink. Overall control may be exercised by providing a variable, intermittent feed to the inking system, such as a transfer roller with adjustable dwell, or by varying the rate of rotation of the roller against the blade: *eg* by a ratchet and pawl system. Lateral control of ink in ducts is usually obtained by adjusting screws which act against the blade to control the gap between it and the roller.

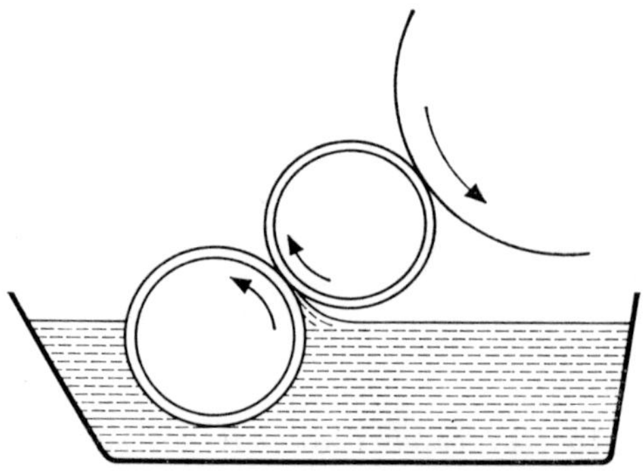

3.3 *Flexographic inking system. The ink container is a simple trough. Metering of the ink film is by adjustment of the gap between the two rollers. This type of inking system is only suitable for inks of low viscosity.*

Remote control of metering is sometimes required on large machines where accessibility may be difficult, or centralised control desirable. This can be achieved in a number of ways such as providing extension rods to the duct screws, motorised electrical control, or in the case of pump systems, extensions of the pipes supplying ink to the distributing systems.

The distribution system. Distribution is necessary to break down the ink structure into a thin, evenly-dispersed film. Generally speaking the more viscous the ink, the greater the number of rollers required to perform this function.

The ink metered out from the duct will be a strip the full width of the rollers but varying from about 10 mm to 50 mm in length and a fraction of a millimetre in thickness according to the amount required to cover the forme. This strip must be split several times and eventually spread out to cover the full forme length of, possibly, one metre. As the layer of ink is squeezed between each pair of rollers it splits into two approximately equal films so that there is a progressive reduction in the thickness of ink on the rollers from one end of the distribution system to the other.

For the ink to pass between two rollers in contact, at least one of them must be resilient, and for this reason distribution systems comprise alternate steel and resilient rollers. Breakdown of the structure of the ink is usually completed by causing some of the rollers to oscillate sideways across the machine to shear the film. If oscillation were not practised with viscous inks there would be a likelihood of the ink pulling itself together to form ribs or strings, with consequent printing troubles. This is particularly relevant to the litho press where the dispersion of water in ink is not based on any affinity between these substances; elimination of oscillation would release moisture to track between the ink body and cause streaking on the printed work. The efficiency of a distribution system to do its job may be measured by the number of ink film splits.

Other factors to be borne in mind include the rigidity of the roller on its stock and the nature of the roller covering materials, which must be compatible with the printing process, possess the required transfer properties, and be of the correct resilience. As the ink is forced into the nip between resilient roller and steel, a hydraulic force is developed which opens up a gap between the two. If the roller stock is not sufficiently rigid or the roller covering material too hard, then one or both of the rollers may bow, allowing a fuller flow of ink through the centre of the system and little at the edges. The same situation may occur if the rollers are set too tightly to each other.

The surface inking system. Inking of the printing surface is carried out by a set of covered rollers which must be capable of receiving ink from the distributing system and laying it as an even film on the image surface. One roller is not enough, as any slight faults would result in imperfect inking. Two or three inking rollers are commonly used and four or even five are not uncommon for good quality printing. The inkers must be supplied with sufficient ink at each impression to completely cover the whole image surface with a uniform layer of ink.

If the supply is not sufficient to provide adequate coverage, then lighter areas may be seen on parts of the print where insufficient ink has been applied. This fault may appear as a general reduction in ink coverage towards the latter end of the print, or a 'ghost' of a previous part of the image where the rollers have not been replenished after completing one revolution.

Ideally the circumference of each inker should be greater than the longest image surface which the roller will have to cover, but in practice, it is rarely possible to design an inking system to this specification. To reduce the possibility of ghosting, rollers should be as large as possible and of differing circumferences. Micrometer adjustments must also be provided for the inkers so that accurate settings can be made between them, the distributing rollers, and the printing surface.

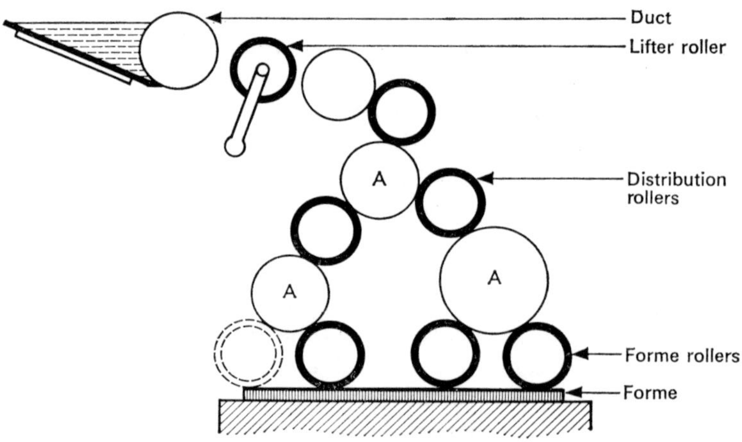

3.4 *Inking system for a cylinder machine. The steel cylinders A are designed to reciprocate laterally across the machine to aid the breakdown of the ink structure. The forme rollers pass over the forme twice before each impression. A fourth forme roller may be included for solid formes.*

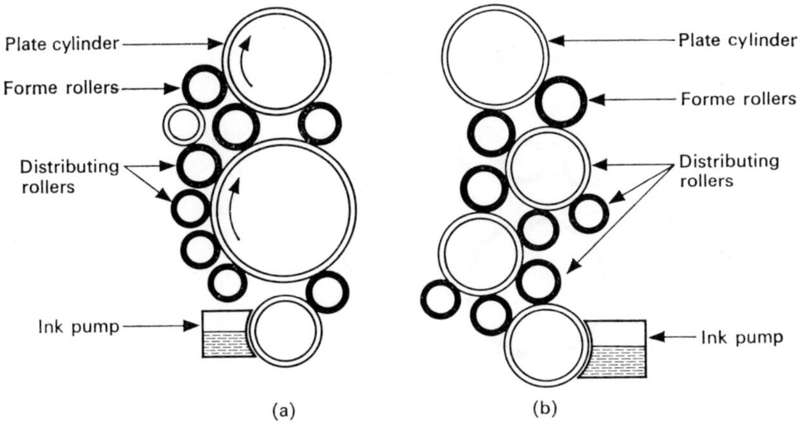

3.5 *Inking systems on publication presses, using an ink pump metering device: (a) Ink is distributed by rollers surrounding a central drum which may reciprocate laterally. This system is not reversible. (b) An alternative arrangement which is reversible.*

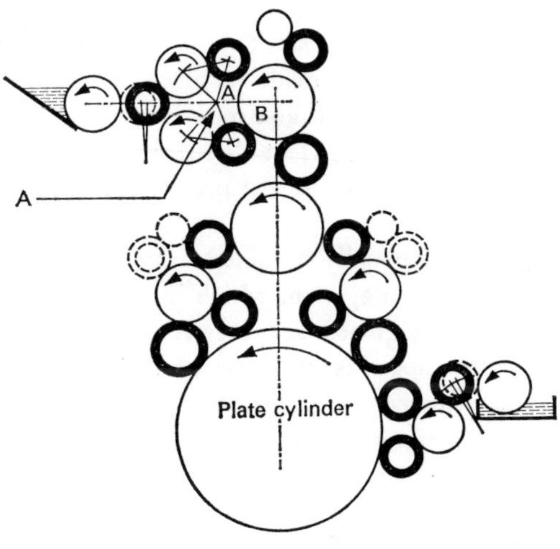

3.6 *Metering and distribution system on an offset lithographic machine, showing conventional forme roller layout coupled with 'progress' inking to pre-mill ink supply to rollers. The section marked A rocks to alternate charging of drum B from either upper or lower feed roller. Conventional damping is also shown.*

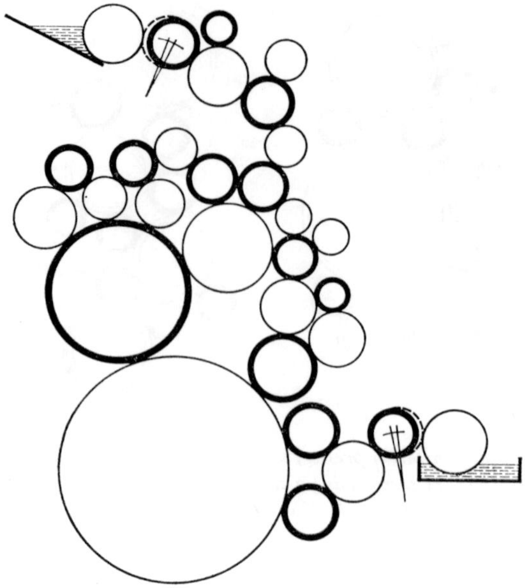

3.7 *Offset litho inking system incorporating a very large forme roller to eliminate ghosting from repeat images.*

Requirements for ideal rolling conditions. To ink an accurate printing surface correctly the inkers should be: 1, made to precise tolerances; 2, concentric; 3, held firmly at the ends to prevent false movement; 4, rigid enough to avoid bowing and whip at high speeds; 5, stable through normal atmospheric and temperature changes; and 6, travelling at precisely the same surface speed as the printing surface.

This last requirement is usually obtained by setting the inkers firmly to a steel roller in the distribution system, which in turn is geared to the bed or plate cylinder. This ensures greater accuracy than gearing the inkers directly to the machine as surface speed is the criterion rather than angular speed. In this way a slight tolerance can be permitted in inker diameter due to wear, atmospheric changes or temperature variations, provided the rollers are otherwise accurate.

Roller settings need to be made with a degree of accuracy depending upon the printing process, the materials being used, and the standard of work required. Even contact and the amount of nip pressure depend upon the following factors:

1 Inadequate contact between roller surfaces will result in ink transfer failure.
2 Excessive contact between roller surface will limit the quantity of ink that can pass through the nip with consequent starving of solid printed work. It may also cause the rollers to bow by exerting more pressure on the ends than the centre.
3 Roller surfaces must be parallel or nip pressure will be greater in the centre than on the ends.
4 Soft rollers transfer ink well, but tend to have a broader nip contact line. They are more prone to dimensional inaccuracy.
5 Harder rollers exert greater milling power and require less contact pressure.
6 Press speed results in nip contact time being inversly proportional to running speed with consequent loss of ink transfer partly compensated by the lowering of viscosity at higher rate of shear. On large machines the whip of the rollers can also cause surface deflection with intermittent loss of ink transfer in the nip centres.
7 The printing process may demand light forme roller contact as with letterpress offset plates.

Rollers are adjusted towards each other, or to the printing surface, either in a straight line or an arc. An arc is better providing that adjustment allows setting to one surface to be made on a radius parallel to the other surface. In this way the mechanics of adjustment to one surface does not affect adjustment to the other. Arc or radial settings are possible by a pivoting action or by use of eccentric adjustments. It is important that eccentric contacting of roller surfaces be made in the right direction or contact will not be parallel to the secondary surface; it is also essential that opposite arcs are not chosen on alternate sides of the machine or roller surfaces will be set out of parallel.

The design problem of eccentrically contacting roller surfaces is the need to get the radius right at the point of contact. Some manufacturers have taken pains to effect this with modern machines and in these cases it is practical to set rollers on the run. Normal practice, however, is to use setting strips of paper, linen or plastic with the machine stationary, testing for an even pull at various points of contact. Sequence of setting will depend on roller accessibility and the mechanics of adjustment. If a mixture of vertical and radial settings are in use, the vertical adjustment should precede the radial since it affects both contact pressures.

Oscillating rollers are gear driven and contact pressures must not be so great that bearings are lifted out of the supporting cups or the gears will be lifted out of mesh and cause irregular inking or gear marking.

Properties of roller covering materials. Where the printing surface is of metal or any other hard substance, it must be inked with a resilient material. The covering of inkers must be resilient enough to avoid damaging the surface, yet sufficiently firm to maintain an accurate shape by resisting distortions due to the ink tack and the pressure of the printing surface as well as the steel roller to which it is set. Roller coverings and similar materials are commonly measured in degrees of 'shore' to determine their relative hardness. The measuring device is called a durometer and employs a scale of 0–100°.

According to machine, type of work and personal preference the hardness of these rollers may vary from 15° to 60° shore. The covering material is also required to be ink receptive and yet to impart the ink readily to the printing surface.

For slow-running letterpress machines the traditional gelatin composition material may still be satisfactory provided its limitations are understood. It has the advantage of being the cheapest covering, being made mainly of glue and glycerine. Its ink transfer properties are ideal, having a resilient, tacky surface which not only transfers ink well, but retains small particles of paper fluff and other foreign matter which would otherwise transfer to the printing surface. The primary disadvantages of this composition are that it is susceptible to temperature and atmosphere changes, will deform permanently if left in contact with a hard surface and has a disturbing habit of swelling until it bursts if used on a high-speed machine. Its lack of resistance to water renders it useless for the lithographic printing process.

Rubber is frequently used as a roller-covering material, being much more stable than composition and practically unaffected by temperature and contact with water. Natural rubber does not have a sealed surface, possesses poor solvent resistance and is too absorbent for general purposes, but it is used on some newspaper machines where colour changes are never required. Synthetic rubbers are less absorbent in varying degrees, and are adversely affected by some common cleaning solvents, notably paraffin, and have no tack whatsoever.

The rubber roller has an advantage in that it can be turned down on a lathe or buffing device to a fine tolerance, and a slightly damaged surface may be restored in this manner. Composition, on the other hand, is cast from a mould and once its sealed surface is damaged it is non-repairable. In letterpress a compromise is to use a rubber roller for stability with a thin surface coating of composition for the ideal surface properties.

Plastic coverings are becoming more widely used as improved versions are being developed. Coverings based on polyurethane or a mixture of selected polyesters and isocyanates can be produced with sealed surfaces, good tack and a variety of hardnesses from 10°–50° shore.

Although atmospheric changes do not affect these plastics, temperature changes do. They are also known to generate heat when running, which may be excessive in an enclosed inking system. The emission of heat in itself is not of prime importance, but the effect of the rise in temperature on the ink can have significant repercussions, bearing in mind that ink viscosity reduces with increased temperature, while drying is accelerated.

Some plastic rollers can be turned down or ground for greater accuracy, though this may affect some of the surface properties, such as tack. Although these rollers are not permanently deformed if left in contact with other surfaces, their tack may be such that damage may occur to them if they are left touching another surface for an extended period and then moved suddenly.

Damping systems on lithographic machines. The most widely used method of plate damping is the conventional one of application from absorbent rollers. Two rollers covered with cotton material contact the plate and are supplied with moisture from a metal distributor, geared and oscillating. Conventional supply to this distributing roller is by intermittent contact with a reciprocating feed roller which picks up moisture from a fountain roller revolving in a reservoir. Alternatively a revolving brush can be used to flick water onto the distributor, leather flappers can do a similar job, or the fountain roller can be eccentric and give an intermittent supply itself. The purpose of alternatives is to maintain a more continuous feed. Conventional dampers synthesise most accurately the hand damping control exercised by earlier lithographers, and provide a scavenging action which removes water droplets from the image, and, with care, helps to remove ink scum from the non-image areas.

Even control of water quantity is essential, and entails fine adjustment of the supply and accurate settings between roller surfaces. The direction of rotation of plate dampers has a bearing on their effectiveness as a damping supply, and extra distributing dampers are sometimes included in the system to allow maximum moisture to the first plate damper.

The number of dampers used and the type of damper covering will control the reservoir capacity of the system and consequently the rapidity with which changes can be made in the damping level. Where control changes are needed within fine limits, cotton sleeve covers of a stockinette type, or paper dampers can be effective. If a slower response will do and supply must be reliable, a heavier grade of covering will give a greater holding capacity for moisture.

As with the ink metering system, there is a need for some lateral adjustment in supply, notably where a sudden change from heavy solid to light image exists across the printing area. Control is best exercised on fountain supply or distributors, and preferred methods are those such as air knives which do not physically scuff the rollers. One alternative method of supply utilises internal feed from within the roller stock, which is made of sintered metal to allow moisture to permeate to the outside cover. This theory has much to commend it as a means of rapid control, but does not permit lateral adjustment.

Moisture need not necessarily be supplied from a separate applicator system. The ink train can be utilised to transport the required level of damping to the plate. Two different approaches are used to put this system into effect. Slow running small offset machines often incorporate a damping feed system metered through the ink supply. Because a milling of ink and water together takes place, wetting agents and salts have to be added to the fountain solution to enable printing to be sharp and clean. Nevertheless a high level of emulsification makes it difficult to print bright solids and the system is not used on larger machines with their greater milling power.

The Dahlgren system offers an alternative to conventional damping more suitable for high speed printing. A simple, accurately metered supply roller feeds damping solution to the first inking roller over the plate. Direction of roller rotation ensures that this film of moisture is applied to the plate before it has a chance to emulsify in the remainder of the inkers. An alcohol enriched supply allows a minimal moisture level and the system allows instant adjustment. The nature of the system makes it effective at high speed but less attractive for low speed equipment.

Spray damping has been employed, but has many disadvantages. Solutions both corrode and contaminate, and it is difficult to maintain a sufficiently finely atomised spray consistently through a run. Moreover there is a disadvantage of water droplet deposits on solids which can only be overcome by using a follow-on roller to pick up superfluous moisture. Similar methods have been used employing flooding of the plate and removing the excess by a doctoring roller. A more recent innovation uses this doctoring roller as the applicator whereby a roller stock is covered with soft rubber possessing a grained surface, to a diameter larger than conventional dampers and the damper ends extended to the bearer rings of the plate cylinder to give a positive drive in the cylinder gap. The supply roller has a conventional covering and damping is improved by adding 20% isopropyl alcohol to water, while refrigeration in the system reduces the evaporation of alcohol.

Gravure

The ink control system in gravure printing can be divided into three basic sections: the container, the method of application and doctoring.

The container in its simplest form is a trough in which the printing cylinder revolves, partly submerged in the ink. The trough must be positioned to catch the residue from the doctor and should be lowered for easy cleaning

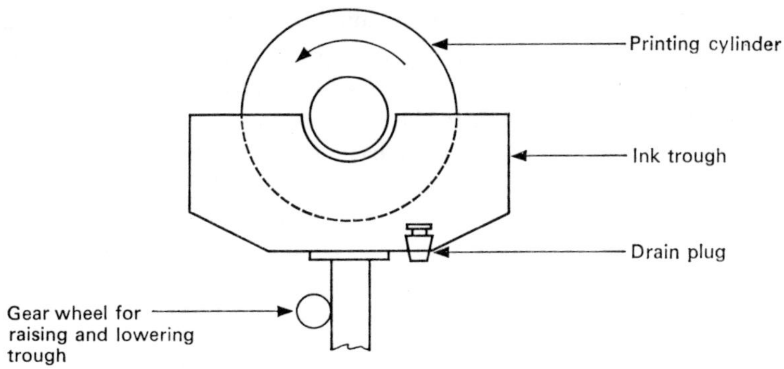

3.8 *A simple gravure ink trough in which the printing cylinder revolves. The printing cylinder performs the function of ink agitation and surplus ink falls back into the trough.*

and emptied by means of a drain-off tap or plug. The main disadvantage of this system is that the cylinder revolving in the trough is the only source of agitation for the ink and may be insufficient to maintain a correct mixture, as well as leading to the ink 'skinning'. For this reason many modern gravure presses have the bulk supply of ink housed in a mobile tank, drum or large storage container with a pumping system to supply the ink to the press. The large storage tanks may be connected directly to the press container or to taps in the pressroom from which ink can be drawn to supply a machine.

Application. The simplest method of application is to have the cylinder revolve in the ink trough. However, this is unsuitable for plates or sleeve cylinders as ink will seep into securing mechanisms and underneath plates. Also on fast-running machines the ink taken up by the cylinder will exert a substantial force against the doctor blade, which will tend to lift it from the image surface. This hydraulic force will be proportional to the angular velocity of the cylinder and the viscosity of the ink.

In other trough systems ink is applied by a roller running in the ink and set to the cylinder. The roller should extend the full width of the cylinder or plate, be robust and accurate in all dimensions and unaffected by gravure solvents. Composition rollers between 15° and 50° shore are suitable, being durable for long periods before they harden and lose shape. Some modified composition rollers last even longer and several plastic coverings are excellent for their stability and solvent resistance. Natural rubber is not suitable as it is attacked by many gravure solvents. Alternative roller coverings are flannelette, upholsterers plush or carpet material sewn round a conventional roller stock or a wooden stock which is segmented to minimise warping. Location to the image is done by teasing strips of paper placed between roller and cylinder and adjusting the settings until a firm overall pressure is obtained. Finer adjustments may be made in the light of running tests.

With ink circulating systems, three basic ways of applying ink to the surface may be used, sometimes in combination with each other. The first is the weir system in which a head of ink is maintained in a trough and a controlled

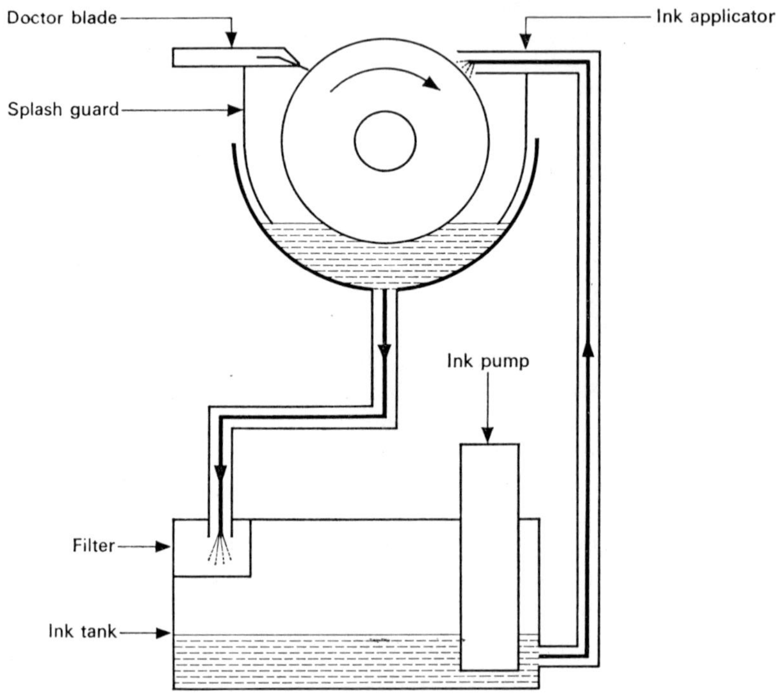

3.9 *A gravure ink circulating system showing spray rail application. Replenishment of the ink is automatic and surplus ink (after doctoring) is returned to the main ink supply for re-circulation.*

overspill fed to the cylinder, the surplus from the doctor blade running back into the trough for recirculation. The second system applies ink by means of a spray rail extending the width of the cylinder, surplus ink falling back through a filtering system for recirculating as in the first method. This method is sometimes combined with a trough in which a head of ink is maintained and the cylinder revolves to augment the initial ink supply. The third system consists of a spray rail and trough which supply ink to the cylinder and a steel or nylon primary blade which meters the ink prior to being wiped by the doctor. This is the principle of the Behringer system. The metering blade gap is in the order of 0·25 mm and must be adjustable as insufficient ink will cause drying of the film on the cylinder prior to doctoring, while an excess will exert pressure against the doctor blade.

Whichever system or combination is used, care must be exercised to ensure an even application of ink along the cylinder, bearing in mind the minimum and maximum cylinder lengths that may be used. Equipment should be designed to keep the ink well mixed, filtered and enclosed to minimise solvent evaporation, and should incorporate pressure controls to ensure the correct supply of ink to the cylinder for any given viscosity or press speed.

Doctoring. The doctor blade rides upon the surface of the printing cylinder and removes the ink from the non-image areas. The success of this wiping operation depends upon the thickness of the blade, the nature of the blade assembly, the angle of the blade edge, contact angle with the cylinder, position on the cylinder and the amount of reciprocation and pressure.

Blades are made from good quality steel in thicknesses ranging from 0·4 mm to 0·1 mm. A typical combination would comprise a 0·2 mm blade backed up by a 0·3 mm supporting blade or possibly two backing blades of 0·15 mm, staggered to provide the necessary flexibility. The best combination of doctor and support blades is the one which will provide maximum wiping efficiency with minimum pressure against the cylinder. Once determined for a machine the combination is rarely changed, although a damaged cylinder which has been repaired by charcoaling and polishing may require a more flexible combination to wipe the area satisfactorily due to the slight depression caused by the repair. Many types of plastic blades have been tested but it has been established that gravure solvents affect most of them, their edges wear quicker than steel and they do not have the flexibility necessary in a blade to enable it to follow the contours of a cylinder turning at high speed.

The blade assembly is held in a frame across the press as near to the printing nip as possible and at as high an angle as practicable. Being near to the nip

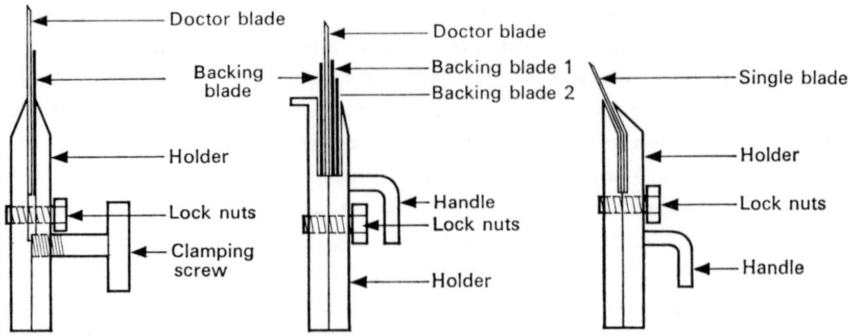

3.10 *Various doctor blade assemblies. The illustrations show different types of doctor blades and holders in current use on gravure machines.*

reduces solvent loss from the ink film before transfer to the substrate and the high angle minimises the area over which the blade edge can wear. It should be possible to adjust the pressure on the blade and the angle of contact to the cylinder. Variations in pressure may be obtained by a simple weighted pulley

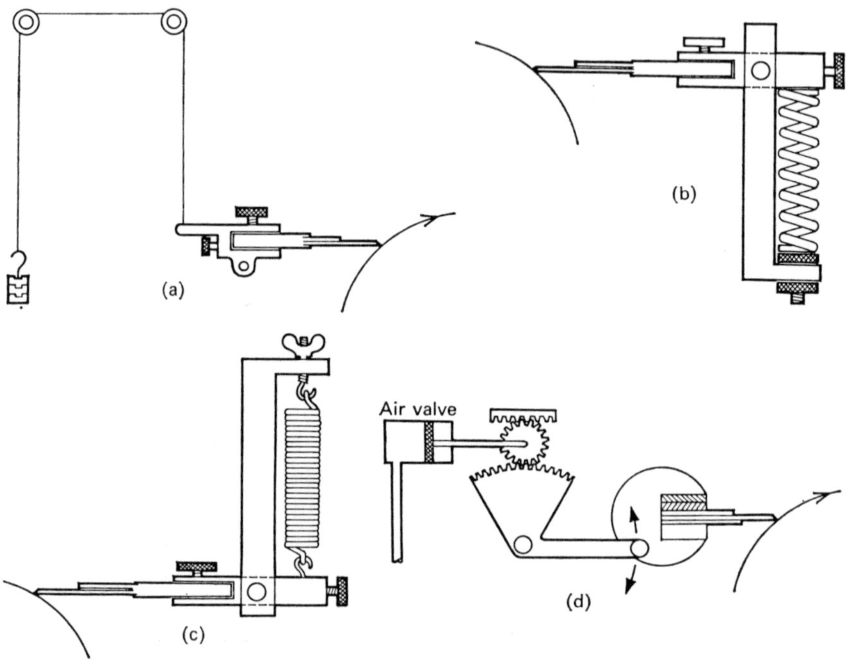

3.11 *Doctor frames: blade pressure adjusted by means of (a) weights (b) compression spring (c) expansion spring (d) pneumatics.*

system or by compression springs incorporated in the doctor assembly. On slower-running machines expansion springs are sometimes used, but there is a risk of vibration, which may cause faults similar to gear marks. On modern presses pneumatic pressure is applied to the doctor which permits pressures to be measured accurately on a dial and re-set after stoppages and machine adjustments.

In non-image areas the blade must strip all the ink from the cylinder, but in image areas the ink film has to be sheared; the angle of the blade edge and its sharpness will therefore affect the quality of the printed result. A sharp blade, ground to an angle of 20° will remove the maximum amount of ink, while a blunt blade, ground to an angle of 90° and with the contact edge slightly rounded by honing and polishing, will leave the maximum amount of ink in the cells. Blade edge angles can therefore help to control print densities where depths of etch vary. The angle of contact between the doctor blade and the cylinder has a similar effect on print densities to the angle of the blade edge and again by adjustment can control image quality. This angle may vary from 15° to 75° with a norm around 45°.

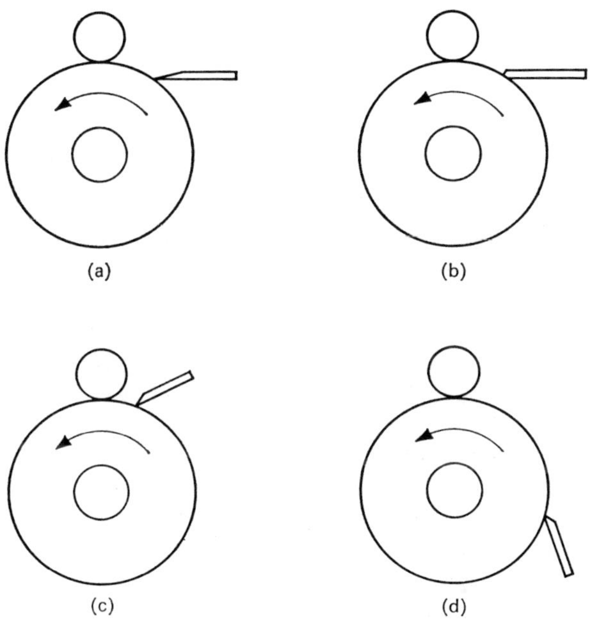

3.12 *Doctoring: (a) A sharp edged blade. (b) A blunt edge. (c) High position of contact. (d) Low position.*

The wiping action of the doctor is normally obtained by setting the blade at an appropriate angle to the cylinder *with* the direction of rotation. It is however possible to set the blade in a reverse direction so that a 'chisel' wipe is obtained and experiments have shown that with this method less pressure is required, blade and cylinder wear is reduced and faults associated with doctoring occur less frequently. The main disadvantage of reverse doctoring is that the blade contacts the cylinder at a position well before the printing nip and the long interval between doctoring and impression gives rise to the problem of ink drying in the cells before transfer.

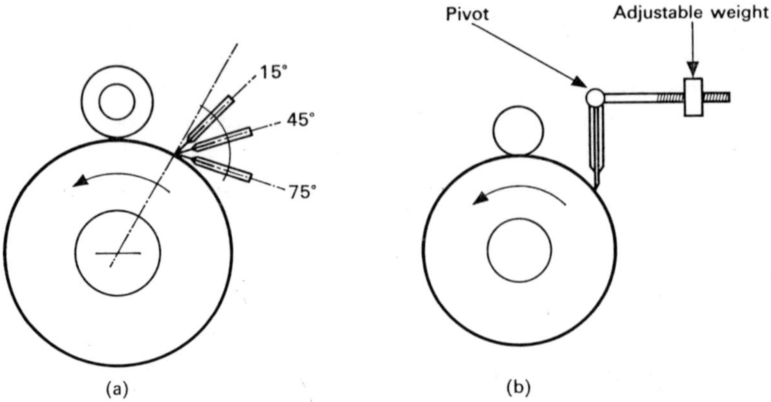

3.13 *Angles of contact for doctor blades: (a) A sharp wipe (15°); a normal wipe (45°) and a flat wipe (75°). (b) Arrangement for reverse doctoring.*

As the doctor blade wears it takes up the shape of its arc of contact with the cylinder, thus losing its prepared edge and becoming less efficient as a knife for shearing the ink. Wear will be greatest where the doctor has been cleaning solids and least where it has been cleaning highlight areas. This results in high and low places along the length of the blade and the situation finally arises where the high parts act as bearer surfaces so that the lower ones fail to clean the cylinder properly. When this stage is reached a newly prepared doctor blade is required.

Reciprocation of the doctor assembly is designed to provide a better means of shearing an ink film than a stationary blade and to extend blade life by spreading the wear along a greater length. Also, in the event of foreign matter lodging under the blade edge, reciprocation will disperse the fine cuts in the cylinder over the area of the doctor traverse, which allows the pressman time to deal with the problem before serious damage is done to the cylinder. Reciprocation is usually adjustable between zero and 88 mm and it is important to ensure that the repeat cycle does not cause cylinder wear, as identified by linear polish marks across the cylinder and a milled edge appearing around the edges of the plate or cylinder.

In gravure printing the ink is metered by the printing surface itself and primary control is therefore the depth of the etch. Some control may be applied by adjusting the blade edge angle and contact angle of the doctor, but the main secondary control is obtained by altering the density of the ink. The ink vehicle is a solvent which holds the binders in solution and the pigment particles in suspension. The proportion of solvent will therefore determine the pigment density in the image cells, and once determined it is essential to maintain the solvent ratio throughout the press run by replacing evaporated solvents with precise quantities of fresh solvent. This control is obtained by measuring the viscosity which gives the most satisfactory result and monitoring at frequent intervals by means of a flow cup, as described earlier in this chapter. An additional check may be obtained by measuring the specific gravity of the ink with a hydrometer.

4. Impression

The nature of pressure required to produce a satisfactory image varies with the printing process used. The factors affecting impression are here dealt with under the general headings of letterpress, lithography and photogravure.

LETTERPRESS

If a flat surface or printing plate is rolled with ink and a sheet of paper laid upon it, very little ink will transfer to the paper unless some pressure is applied. The pressure required will vary according to the properties of the paper, the ink and the printing surface.

For instance, a free-flowing ink of low viscosity will transfer readily from a non-receptive surface to a tacky, absorbent or otherwise receptive material under a pressure of less than 100 kNm^2. Such ideal conditions are rarely obtained in practice because one would have the problem of laying ink on the non-receptive surface in the first instance, and the material upon which the print is made cannot always be completely ink-receptive.

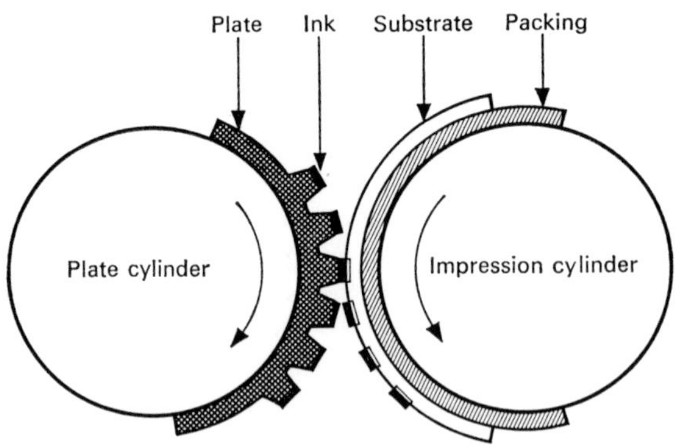

4.1 *The impression 'sandwich'. Ink in selected areas determined by the printing plate is presented to the substrate under pressure. A compressing force is developed in the nip area in which all surfaces are involved, eg plate, ink, substrate and packing. There are five inter-surfaces: 1, plate cylinder/plate. 2, relief surface/ink. 3, ink/substrate. 4, substrate/packing. 5, packing/impression cylinder.*

For these reasons pressures of several hundred kNm^2 are more commonly required, and where transfer conditions are far from the ideal, the pressure may exceed one meganewton per square metre.

There are three distinct, but related, phases in making an inked impression: 1, the bringing together of the substrate and the printing surface; 2, the application of pressure between the two, making an ink 'sandwich'; and 3, separation of the surfaces with a resultant transfer of ink to the substrate.

Pressure is necessary in printing in order to bring the surfaces into sufficiently close contact for ink transfer to take place. If every component in an impression system were rigid, *ie* made of metal, then the force required to ensure overall contact would be of a very high order and would depend to a large extent upon the accuracy of the surfaces.

For example, if two flat metal surfaces are brought together and required to make overall contact then unless the two surfaces are perfectly level the forces bringing them together would need to be sufficiently high to ensure some compression. To obtain a compression of as little as one micrometre in most metals would require pressures of the order of 10–20 GN/m^2.

The strain on a machine designed to perform such a function would obviously be enormous and would demand an extremely robust construction even if the tolerance of one micrometre were practicable. For this reason, and others, a compressible element is usually introduced into the printing system. Some assistance may be obtained in this respect from the printing stock if it is, for example, a bulky 'antique' paper, but as a rule the material is too thin for its compression to be significant.

The compressible element may be either the printing surface itself, a rubber or plastic forme, or the impressing surface may be covered with a compressible material, the paper 'dressing' or a patent blanket system. Whatever the compressible material and wherever it is situated, it should also be elastic so that it recovers from impression before the next print is made.

Incorporating a compressible material brings a number of benefits: 1, the pressure required from the machine is lowered, therefore reducing the need for excessive strength in the impression mechanism; 2, wear on the machine and the printing plates is reduced as well as the possibility of damage due to crumpled sheets, etc; and 3, a reasonable tolerance may be determined for printing plates and other materials, in the order of hundredths of a millimetre instead of thousandths.

Impression between plane surfaces

If a force is applied to a flat surface in contact with another flat surface, an equal and opposite counterforce will be developed in the second. Provided the two planes in contact are accurate and of identical dimensions, the forces will be distributed equally over the whole area, resulting in an even, overall pressure.

When an impression is made on a platen machine there are several planes to be considered; the printing surface, the two sides of the ink film, the two sides of the printing material and the impressing surface. The coming together of the platen and forme exerts a force on the ink/paper sandwich. If the pressure developed is too little, ink transfer will not be satisfactory, while too great a pressure may result in indentation of the stock and the platen packing, as well as causing undue wear on the printing surface and in the machine. Between these two extremes there is usually a fairly wide range of satisfactory printing pressures.

Where the image to be printed is something other than a regular solid shape, possibly an assembly of small, irregular shapes such as type matter or a halftone illustration, then the force applied will still be distributed over the whole surface. But because a constant force is being applied to disproportionate areas, the pressures will vary inversely according to the areas.

This principle can be illustrated by the printing of a simple forme in two halves. One side is a solid area 100 mm square, while the other side is an open box 100 mm square made up of one millimetre rules. The solid will have a printing area of 10 000 mm^2 and the box will have an actual printing surface area of only 400 mm^2.

If the forme is inked and a force of 1 kN applied then a pressure of 100 kNm^2 will develop in the solid area, which is hardly enough to make a print under average circumstances. At the same time a pressure of 2·5 GN/m^2 will be exterted on the rules in the box area, which will almost certainly cause identation of most printing stocks. It is obvious that in such a circumstance some method of pressure equalisation will be necessary.

Pressure equalisation

It has already been stated that the introduction of a compressible, but elastic element in the printing system will allow a range of pressures to be used so that ink transfer can be effected without incurring faults such as indentation or image distortion. However, where the size variation of the forme elements requires pressure equalisation beyond this tolerance a further operation is necessary.

There are two basic methods of equalising pressure on platen machines. One is to vary slightly the heights of the printing elements in proportion to their areas so that the applied force is evenly distributed, and the other is to adjust the counterforce from the platen by means of overlays. In either case

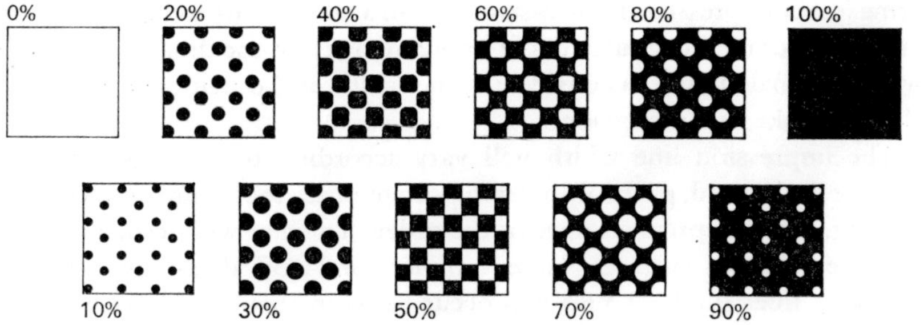

(a)

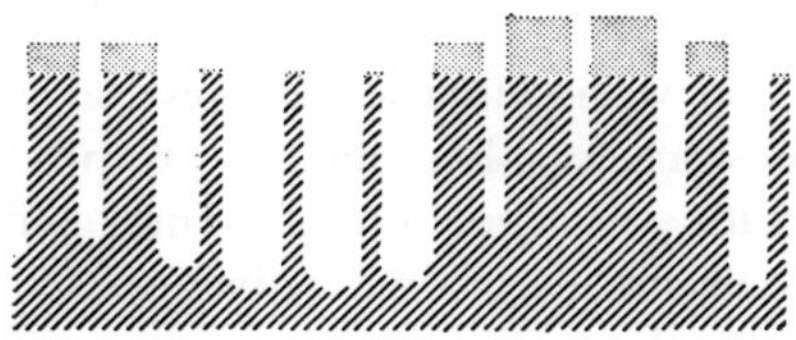

(b)

4.2 *Halftone printing. (a) Various tones are obtained by printing dots of differing sizes. (b) Shadows need more printing pressure than highlights for perfect reproduction in letterpress printing. The differential pressure adjustment is obtained in modern plates by slight variations in plate height, proportional to the printing areas as shown in this illustration in exaggerated form. The overall plate height differences between highlights and solids will probably not exceed 0.05mm.*

the adjustment will only be in the order of a few hundreths of a millimetre between minimum and maximum heights.

Height variation may be achieved by applying a thin, paper 'interlay' between a plate and its base, or an 'underlay' below the forme. Alternatively, printing plates or type may be produced with built-in height variations. Several modern platemaking systems do in fact have in-built height variations according to the tonal value.

Overlays can be simple hand-cut 'patches' which would suffice for the basic forme previously mentioned, or they may be obtained by more elaborate patent systems of which there are several on the market. These mechanical overlay systems will automatically produce, from a proof of the job, the subtle height variations required for the fine detail of halftone work.

Cylindrical impression

When an impression is made by a cylinder, most of the basic principles relating to platen impression still apply. The most important difference is that the forme is not impressed at one instant, but in a continuous rolling action with only a strip of the image area under impression at any one time. Here we have two additional factors to consider: 1, the width of the strip; and 2, the time taken to make the impression.

The impression line width will vary according to the type of forme, machine design and, probably most important of all, the nature of the cylinder dressing. Wide impression lines of the order of 10 mm will occur with compressible printing surfaces, large cylinders and soft dressings, while the narrowest lines (under 5 mm) will occur between small cylinders with hard packings and rigid formes. The impression line is also influenced by the accuracy of the forme and by the pressure developed by the cylinder in order to make a satisfactory print.

It is also obvious that with cylindrical impression the force required to produce a print can be considerably reduced; *eg* to print a solid 100 mm $\times$ 100 mm on a platen, with a firm pressure of 1 Nmm^2 will require a force of 10 000 newtons, but the same image on a flatbed cylinder or rotary machine with a 'nip' of 5 mm will only require 500 newtons to produce the same printing pressure.

Time taken to make an impression is not significant when there are only slight variations in printing speeds. On a flatbed machine with a speed range between 1 000 and 5 000 impressions per hour the difference in impression between minimum and maximum speeds would be hardly measurable. But if a newspaper machine were set to a good impression at its slowest speed of say 200 rph and then accelerated to 20 000 rph the reduced pressure dwell would quite possibly mean a complete loss of impression.

Pre-tension

Rolling impression does not mean that a machine need not be strongly built. A force of several kilonewtons is still required to print a solid subject from a metal printing surface on a rotary machine. If the construction is not strong enough to produce the required pressure then the cylinders will be forced apart by the forme. This will give rise to the fault of 'guttering' as the cylinder alternately rolls over the forme elements and drops in the gaps between.

The weight of a cylinder alone will not produce sufficient pressure to print solids. Therefore the cylinders are squeezed together, usually by a mechanical adjustment, which maintains a state of tension between the cylinders. This force acts through the bearers on the ends of the cylinders, resulting in a stress which is called 'pre-tension'. The amount of pre-tension is usually adjustable and may vary in degree from one type of machine to another. It can be measured with a suitable strain gauge.

The counterforce required to move the cylinders apart a distance of 0·1 mm is used as a basis to compare the 'stability' of different machines. The figure resulting from this test is referred to as the 'stability index' of the machine in question. The stability indices for most flatbed cylinder and rotary machines lie between 10 and 50 kilonewtons per 0·1 mm change in gap.

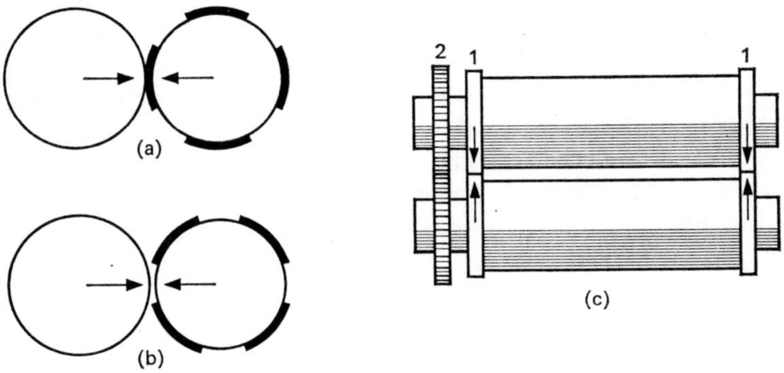

4.3 *Pre-tension. (a) A plate cylinder and impression cylinder rolling in contact are squeezed together so that stress occurs at the line where they meet, producing the necessary pressure required for printing. (b) The cylinders come periodically under conditions of stress and non-stress according to the positions of the relief images on the plate cylinder. In the non-stress phases the cylinders will move towards each other, and will then be forced apart when the relief surface enters the nip region. This situation results in indentation at the edges of the image and rapid wear of plates and impressing surfaces. (c) To counteract the above situation, bearers (1) are fitted to cylinders so that they can be squeezed together in a permanent state of stress, regardless of whether the relief surface is in the nip region or not. The bearer diameter is the same as the pitch circle of the gears (2) and also serves as a positive reference for correct caliper of plates and packing.*

Letterpress blankets

A blanket is used in letterpress printing instead of the traditional paper dressing to reduce or eliminate the need for makeready. Makeready in this context refers to the adjustment of printing pressures by the machine printer, and there are two reasons for its practice; one is where forme components are not within the established tolerance and must therefore be adjusted to correct height; the other is where area variations in the printing image are sufficiently diverse to require a pressure equalisation procedure.

It has been stated that the dressing is needed to allow printing to be satisfactory between a range of pressures. The purpose of a blanket is to enable a wider latitude of pressures so that ink will be transferred without indentation of substrate or distortion of the printing image.

The ideal blanket should be accurate in caliper to within 0·01 mm and should have a firm surface to enable a sharp outline of the image to be obtained, but a resilient sub-surface to allow some compression to take place. It should compress without displacement as far as possible and should not become measurably denser under compression. These last two requirements are not compatible and it is usual to find that resilient materials under pressure experience varying degrees of displacement and compression.

The blanket should not stretch significantly when tightened around its cylinder as this will amost certainly mean a change in caliper. In some instances unequal caliper changes due to stretching have been known to amount to variations of more than 0·1 mm. Rapid recovery from compression is also essential so that the blanket is back to full caliper before the next impression. Available time intervals for recovery vary from one second or more on a flatbed machine to less than one fiftieth of a second on a common-impression reel-fed rotary.

The original 'automatic' blanket was developed in the newspaper printing industry, where there is never the time available for the practice of make-ready. This was a felt-based blanket with a bonded surface of a nitro-cellulose plastic, used in conjunction with a rubber underblanket. This dressing is still in use for newspaper printing but in other sections of the industry different combinations of hard surface materials with soft backings are frequently used. Typical combinations are cork backing with a polyester surface, and micro-cellular rubber with a surface layer of PVC.

LITHOGRAPHY

In offset machines the intermediate blanket cylinder takes over from the plate cylinder as the image impressing surface. The cylinders do not rotate together in perfect osculation but are squeezed together so that an area of contact is established rather than a theoretical 'hair line'. This is because a certain amount of pressure is required to effect ink transfer from one surface to another, as in letterpress printing. However, in lithographic printing the pressure is applied evenly across the nip line and is therefore always directly proportional to the force applied between the two cylinders and the area of contact between them.

Forms of printing pressure

For practical purposes the printing pressure is the squeeze between the cylinders as governed by the impression adjustment system. There are three methods of pressure setting in common use: bearer contact, bearer clearance and surface setting.

Bearer contact design requires accurately machined bearers which revolve in contact at the pitch circle. This provides continuous traction and consequently the minimum vibration and wear. The method of setting is to take off about 0·1 mm and dab a thin film of ink on the blanket cylinder bearers approximately 100 mm from the gripper edge, then adjust the settings slowly until ink transfers to the opposing bearers. A slight additional pressure may sometimes be advantageous, but an excessive increase could result in bowing of the contacting surfaces. Each cylinder will be dressed to fill the recesses, plus the additional amount required to obtain the necessary pressure. The sheet to be printed should be considered as part of the impression cylinder dressing with this system and a packing provided which can be adjusted according to the caliper of the substrate.

Bearer clearance machines are usually set up by using gauges between unclothed cylinders. On some presses the position is pinned once the machine is set up and further adjustment may be possible on compensating scales which indicate variations from the norm. Each cylinder is usually packed out to 0·05 mm above the recess so that an interference pressure of 0·1 mm is obtained between a pair of cylinders. Allowance must be made for different calipers of stock.

Surface setting is the more usual method of establishing impression. This can be done by the adjustment of quadrants on the eccentrics which actuate the impression movement. Quadrants can be set to zero with gauges between unclad cylinders and both sides set independently. When the quadrants are locked a single, calibrated adjustment can be arranged to set impression on both sides of the machine to allow for variations in stock caliper.

Impression trip

The cylinders need only be in contact when printing. At other times their relative positions should be adjusted so that they are sufficiently apart to avoid printing on the impression cylinder and excessive build up of ink and water during non-printing operations. Similarly when the machine is stationary the cylinders should be out of contact to prevent the possibility of flats developing between contacting, compressible surfaces and gum, ink and water from dwelling under pressure to cause streaks.

The impression trip is usually controlled by a limited rotation of the eccentric journals in which the cylinder shafts are housed, and the operation may be effected manually or automatically. Manual operation is usually done by disengagement at some point through the trip linkage, which will also lift the rollers and stop the ink feed. The action must be rapid so that the cylinders are separated between the trail edge of the plate and the lead edge to avoid printing on the impression cylinder. Automatic operation is essential on multi-colour machines to ensure that cylinders are tripped in or out in correct sequence.

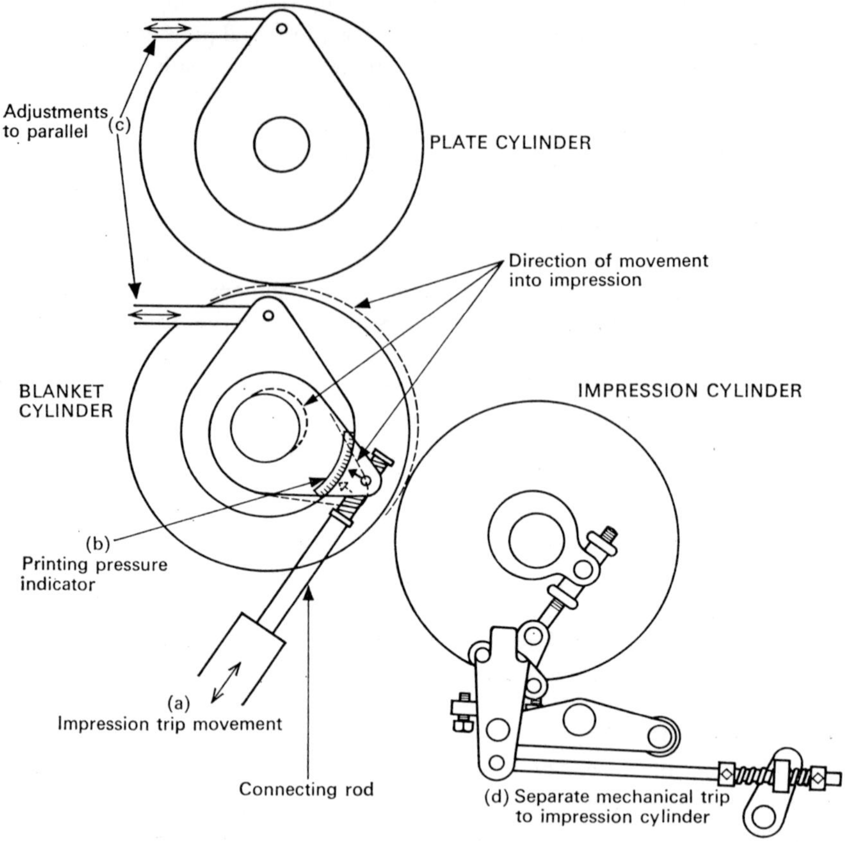

4.4 *Conventional offset trip arrangement.* (*a*) *Normal pressure disengagement to intermediate cylinder may be pneumatic, electrical, or mechanical, and may have facilities for setting to parallel built-in to actuating arms each side of the press.* (*b*) *Adjustment to pressure for different calipers of stock, usually to moving connecting rod linked to setting scale.* (*c*) *Parallel adjustment to blanket cylinder may be to a fixed point or move with trip. Alternative plate parallel adjustment is usually to a fixed point as plate cylinder rarely moves with trip.* (*d*) *On tin-printing machines the second eccentric trip is effected on the impression cylinder to enable independant action between plate/blanket and blanket/impression. Actuation may be connected to misfeed trips either hydraulically, or mechanically as in the diagram, which illustrates the walking beam and roller principle.*

The blanket cylinder is usually considered the most convenient one to trip out of control, as it can be adjusted in relation to the other two main cylinders by one movement and does not affect paper feed or inker and damper settings. This is not always possible, however, as explained in chapter 1. For instance, on machines for printing on metal it is necessary to pre-charge the blanket with the correct printing film of ink before going on to impression cylinders to avoid printing on the latter. This is accomplished by having separate eccentric journals on the two cylinders, and the separation of impression cylinder tripping also helps to prevent smashes with misfed metal sheets.

The printing nip

As lithography involves the principle of complete impression and the presence of moisture on the printing image, it is apparent that both moisture and squeeze variations will affect the print dimensions. Fortunately, almost total reliance on flexible plates enables adjustment to be made in print size in the machine direction, *ie* around the cylinder circumference. Basically the theory involves correction of dimensional change occurring in the impression nip by a compensating adjustment to the radius of the printing cylinder, this being altered by reducing or increasing the packing under the machine plate. It follows that the image will occupy a smaller or larger proportion of the plate cylinder circumference depending upon whether this has been increased or decreased. The particular machine pressure setting system may also demand a compensating adjustment to blanket packing to maintain correct interference pressures in each nip.

It can be seen that removing packing from under the printing plate and, if necessary placing it under the blanket, will lengthen the print. Alternatively, more packing under the plate and less under the blanket will shorten the print. The following example illustrates the point:

It is required to increase the print length; packing calipering 0·05 mm is therefore removed from under the plate and inserted under the blanket. The accumulated change in the diameters of the two cylinders is now 0·2 mm. The difference in each circumference will be 0·1 mm $\times$ π (3·1416) and the total difference between the two cylinders is 0·2 mm $\times$ π = 0·628 mm approximately. If the design covers half the cylinder circumference, then the image is lengthened 0·3 mm on the blanket.

It is important to note that only circumferential corrections are possible. Adjustment across the plate cylinder is confined to cutting the machine plate down a suitable vertical margin and plating-up the separate parts. Opportunities to do this are rare, and the requirement is normally avoided by running paper to be printed long grain. This means the paper fibres will lie across the machine (this usually being the longest way of the sheet) and the minimal dimensional changes will occur in this direction as a result of moisture pick-up. The maximum change will then occur around the cylinder and can be compensated for. Web machines are excluded in this context since they always run with the grain in the machine direction.

Theoretically, packing changes to the intermediate blanket cylinder do not alter printing size since a gain on the blanket becomes an equivalent loss at impression. In practice some change does occur, because of the resultant variation in comparative cylinder surface speeds and the dependence of paper substrate on surface friction to control its passage through the impression nip. Blanket and impression surface speeds should be the same to get a true impression of the image printed and in order to obtain this ideal, some bearer contact machines have packing facilities on the impression cylinder to maintain full control of peripheral speeds whatever the substrate printed.

In practice a tacky blanket surface drives the sheet through impression and a certain amount of change in print size can be effected by blanket packing. On flatbed proving presses it is the only way a size change can be attempted other than the obvious one of processing new plates. In metal printing, however, the surface speed of the blanket must be constant since sheets of tinplate ride through without gripper control; packing variations will therefore influence sheet drive through the impression nip.

Fragile paper stocks are susceptible to tearing or creasing if the blanket should race or drag them through the impression. Web machines again provide the exception since tension of the web through the impression nip can be controlled and print size altered by changes in blanket cylinder diameter.

The race or drag caused by altering the circumferences of the plate, blanket and impression cylinder can set up mechanical forces which in turn may cause impressional problems. Firstly, there is the effect of tangential forces resulting from two contacting gear-driven cylinders which differ in diameter; secondly there are compression forces resulting from the resistance of a resilient covered cylinder. Accuracy of cylinder drive is essential for clean, sharp, dimensionally-correct image reproduction, and the importance of gearing in this context is emphasised in chapter 1. The lead edge of blanket impression is vulnerable to the effects of backlash where gear play exists, since the contacting cylinder surfaces will tend to take over drive control from slack gears. This will result in print slur at the gripper edge of the sheet unless facilities are available to take up this backlash by means of adjustable split gears or other means. The extra gearing may exist as a segment at the lead edge, or the whole gear may be duplicated by adjustable rigid or sprung segments meshing together.

If two cylinders of different diameter are allowed to rotate freely in contact, a 360° rotation of the smaller cylinder will not complete a 360° rotation of the larger one since the surface is of greater dimension. When identical axial rotation is forced on such contacting cylinders by gear drive and one of the cylinder surfaces is a resilient blanket or roller, a build-up of blanket occurs behind the contact line if the blanket cylinder is larger, or in advance of the contact line if the blanket cylinder is smaller. The distortion pattern is into the bulge and outwards to the trail edge until release occurs in the cylinder gap. Where the build up occurs in front of the nip at impression, paper will be chased in the centre and will draw in and crease at the trail edge. The more common build-up behind the contact line elongates the paper, and spreads the rear edge under impression. The subsequent incomplete contraction of the stock upon release narrows the image dimension increasingly with each printing that has to be made. Register discrepancies are likely to occur where a number of colours have to be superimposed and sheet size is large.

Machine manufacturers of large presses have tried to overcome the register

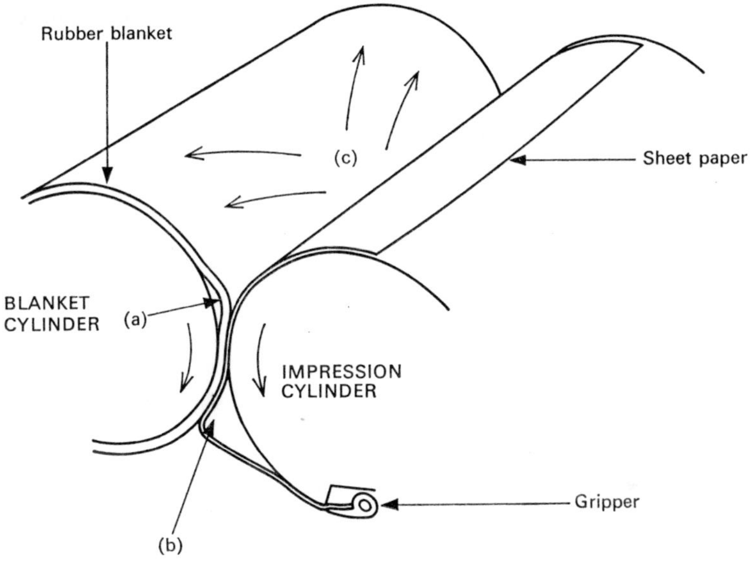

4.5 *Impression forces on an offset blanket. (a) Blanket bulge in front of contact line where blanket surface is larger. Compression occurs at the cylinder nip. (b) Paper is released from the blanket beyond the impression nip. (c) Pattern of blanket distortion under compression. Bulge in front of contact line causes trail edge of paper to spread with blanket flow. The converse condition causes flow inwards and consequent creasing.*

hazard resulting from sheet distortion through impression by curving the lead edge of the sheet through its first impression so that the outer sides are in advance of the centre and the trail edge fully stretched. This has the effect of increasing the contraction of the image after printing so that it is easier to fit on subsequent printings, when the lead edge can be straightened to lessen the degree of distortion. One method employs a feed-roll system of stroking the sheet into register position against a number of adjustable stops so that the sheet buckles slightly before the grippers close. This method of forcefeeding gives a control at the feed end of the machine but does not allow for changes between units on a multi-colour press. Another method has been evolved of bowing the lead edge by angling the transfer gripper rails from the centre, a system which is suitable for the first unit and between successive units. The gripper mounting rails pivot from the centre so that a greater or diminished movement can be accomplished in comparison to the sides. Adjustment can accordingly be made on the run within the limits demanded by the need to prevent sheet creasing. Pressures should be kept as light as possible to reduce the problem of sheet slippage.

The resilience of the blanket cylinder increases its effective circumference

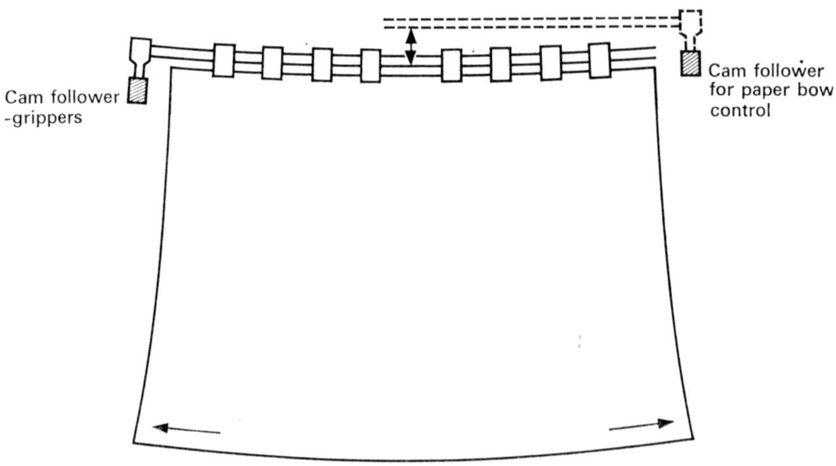

4.6 *Control of paper bowing to obtain register on large offset presses. Large sheets spread through impression and do not completely contract, so that registration is difficult on multi-colour work. Exaggerating the spread by concaving the lead edge on the first printing enables stretching of the trail edge to be relaxed on the subsequent units so that a fit may be obtained. Gripper trolleys can be eccentrically controlled from the centre of the shaft to control the shape of the lead edge through the printing cycle.*

under impression. This means that machine cylinders traditionally constructed with equal perimeters do not necessarily result in equal surface speeds. Manufacturers take this into consideration when designing presses to incorporate aspects of a true rolling system. This means that the blanket perimeter will be slightly smaller than that of the plate to give equal peripheral speeds under normal compression. Similarly the impression cylinder must also allow for the printing stock, which increases the diameter of this cylinder. Combination of true rolling cylinder design between plate and blanket, coupled with use of the smaller action angle (described in chapter 1) when gear cutting, makes it practical to adjust the plate packing without need to alter the packing of the blanket.

The length of the first printed colour tends to increase with expansion of fresh paper, due to moisture pick-up from the lithographic process. It is normal to overpack this plate to reduce print-length so that subsequent colours will fit with progressively reduced plate packings. Where cylinders are designed on the equal-size principle, the effect on later printings is to increase the build-up of blanket after the nip since its peripheral speed is comparatively greater than that of the plate cylinder. With true rolling this difficulty is minimised.

Printing conditions at impression are dependent upon the chemical and physical nature of the blanket, ink, and paper, and involve tensile forces exerted by ink transfer affecting the way the substrate strips off the blanket. Tack of blanket and ink causes the substrate to cling to the blanket and be released beyond the impression nip. In consequence a volume of air is contained between the point where the stock leaves the blanket, and the impression cylinder. The greater this volume becomes through stock elasticity or excessive tack, the more likely it is that the substrate will be elongated, tear, or release to follow a new course round the blanket. Variation in the areas of print across the contact line will vary the strip-off conditions and cause the paper to distort across the press to the detriment of sheet feed to the next unit or machine pass.

Low tack conditions are essential to high speed performance since paper stock is more likely to fracture under the increased tensile forces at high surface speeds. The point of release is taken further through the nip as speed is increased unless a proportionate decrease in tack or acceleration of solvent absorption is made. Pressure will be adequate for complete contact with the stock and cannot be reduced for high speed printing as the moment of impression is shorter. Absorbency characteristics of paper stocks and their surface strength become more important under these conditions and pre-testing is essential to maintain a consistent standard.

Offset blankets

Apart from suitable printing characteristics of receiving and transferring ink, offset blankets must have an even surface without caliper variation, resist stretch under tension, and have chemical and physical tolerance to oils, solvents and water. They are manufactured from laminates of resilient rubbers with natural or synthetic canvasses. Synthetic nylon and fibre-glass backings improve stretch resistance and synthetic rubbers improve solvent tolerance. Softer rubbers have a better affinity for ink, but the need to obtain a smooth even surface without excessive ink tack demands a hardness combined with resilience which complicates blanket manufacture.

Rubber is not truly compressible since impression downwards into solid composition results in an outward displacement equal to compression. The effect of this is to produce squash in the nip and enlargement of the image at the edges. Blankets which are compressible are now made by retaining air pockets in the surface lamination, or by using compressible media such as cork as underpacking. The effectiveness of such blankets is most marked when compressibility is introduced into the surface lamination, but it should be noted that the theory of true rolling becomes less relevant and some change in packing ratios must be made.

GRAVURE

Factors affecting pressure required

The amount of pressure required to transfer ink from the gravure image will vary according to 1, the properties of the substrate; 2, the sheet or web-path; 3, the type of image; 4, the nature of the ink; and 5, the design of the impression system.

The substrate is probably the greatest variable. For instance, although an absorbent paper stock will readily accept the fluid inks used in gravure printing, sufficient pressure must be exerted to compress the paper slightly in order to deposit ink in the small pits and recesses that are always present in a fibrous material. Plastic films, on the other hand, have smooth even surfaces and it is only necessary to lay the ink on the material with sufficient pressure to ensure transfer. As in lithographic printing, pressure is applied evenly along the length of the image and varied by adjusting the relationship between the printing cylinder and its impressing surface. The nip line resulting from the pressure between the two surfaces can vary from 13 mm to 20 mm and some typical pressures for a wide range of substrate are as follows:

Polythene and similar extensible films of about 30 micrometres: 1·5 to 2 newtons per linear millimetre.

Cellophane and cellulose-type films of about 30 g/m²: 6·5 to 8 newtons per linear millimetre.

Papers of about 80 g/m²: 9·5 to 12 newtons per linear millimetre.

Carton board of about 250 g/m²: 15 to 25 newtons per linear millimetre.

The sheet or web path. On sheet-fed machines the paper passes around the impression cylinder to meet the image at the printing nip and is taken sharply away by the impression cylinder after printing. The situation is not the same in reel-fed machines as the web path is largely independent of the impressing surface. In this case the angle of entry of the web towards the printing nip will determine the position at which it will come into contact with the image. A low-angled entry will cause the substrate to meet the image earlier than a high-angled one, and it follows that the angle at which the web leaves the nip has similar implications.

The image will not print accurately if the impression is not adequate. A conventional square cell may sometimes be seen under magnification to print as an outline, with no ink transferring from the centre of the cell. This fault is referred to as 'cell wall printing'. A similar fault with invert halftone images can be seen as a ring of ink transferring from the edge of a cell only. In this instance the fault is known as 'doughnutting'. These faults may be aggravated where the compatibility of ink and substrate is poor and where static charges in the substrate repel the ink in the cells. The problem of unwanted static charges in ink and substrate can sometimes be turned to advantage by inducing charges of opposite poles so that mutual attraction occurs.

The ink should have a preferential affinity for the substrate rather than the image metal and must have a high internal cohesion so that under impression the whole of the ink from each cell is transferred. If cohesion is low the ink may shear, with only part of the film transferring to the substrate and the balance remaining in the cells. This situation will encourage the likelihood of residual ink drying in the image and affecting print quality. As gravure inks are temperature sensitive, high temperatures in the image metal or the substrate may cause accelerated migration of solvent molecules within the ink, which may again cause drying in the cells and affect the amount of ink transferred. A further factor affecting the print from the gravure image is the traverse of the doctor blade. This displaces the ink within the cells in different ways according to the relevant motions of the blade and the image and means that a highlight cell may not print as a perfect square due to the ink being displaced in the cell immediately prior to impression.

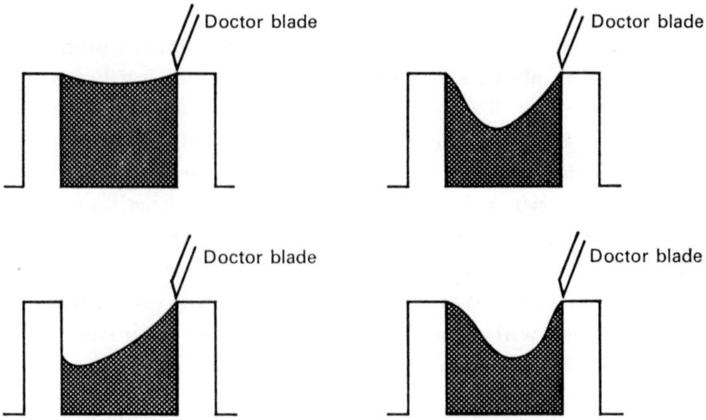

4.7 *Theoretical profiles of ink in cells after doctoring. The behaviour of a liquid ink upon shearing must be qualified by its behaviour in a confined space such as a photogravure cell. Some of the possible post-shearing effects are shown in this illustration.*

Impression systems – sheet-fed

The impression is provided on sheet-fed machines by a cylinder covered with a rubber blanket, built up to correct periphery by the insertion of additional temporary packing. The diameters of impression cylinders are related to the image cylinders which the machines are designed to accommodate. Where a machine is designed to take a range of image cylinders there must be appropriate provision for adjustment between the cylinders and this is usually integrated with the impression setting mechanisms. Bearers are not provided on either image or impression cylinders in gravure printing. Impression setting is carried out by placing strips of paper between the cylinders and adjusting the setting mechanisms while teasing the strips. The adjustments are finalised by reference to the printed image.

The temporary packing should extend around the cylinder precisely to the back edge of the sheet. When the packing sheets are pasted into the cylinder they should be staggered by about 1 mm to ensure that the impression cylinder rolls on to impression rather than bumping on. Bumping sets up a vibration in the framework of the press which transmits to the doctor blade, causing a pressure variation which affects the reproduction across the cylinder. This is known as a 'kick' or 'bar mark' and if allowed to continue will cause premature wear of the image. Excess impression will aggravate this problem and worn cylinder bearings may contribute to the oversetting of pressure.

If the temporary packing should be larger than the job sheet, the rubber blanket will receive a microscopic ink film from the surface of the printing cylinder in the non-image areas, which occurs when the doctor blade begins to wear. This film builds up on the rubber blanket until it is re-deposited on the image cylinder where it travels to the doctor blade and nicks it. As the blade reciprocates, lines of ink arising from the nicked blade will print on the edge of the substrate.

A correctly packed impression cylinder will run at the same surface speed as the printing cylinder, but where the packing ends the periphery of the impression cylinder decreases; therefore at this point there is a difference in surface speeds resulting in a slip between the two surfaces which wears the face of the image cylinder. This will eventually produce a tone mark and the aim should be to cut the packing to ensure that this mark occurs in the trim area.

If an examination of the printed image shows that insufficient ink is being transferred from the cells, selective pressure in the form of overlays can be applied. These are sheets one millimetre larger all round than the images, mounted and registered accurately beneath the rubber blanket. They exert pressure on the image areas with the result that more ink is removed from the cells and less pressure is applied to non-image areas, which will consequently run cleaner for longer periods. The overlays compress the packing and there is no need to reduce the packing thickness when they are mounted.

Before the image is etched all colours are accurately laid to the key cylinder and the main problems of register on sheet-fed machines are those of substrate stability and machine settings. If, however, the pressure is set unevenly this can lead to register variations throughout the press run. Establishing register requires setting the front lays to grip the correct amount of paper then relaxing bolts or removing a gear in order to position the image in relation to the substrate before re-locking or re-engaging the gear. Final register and running adjustments can then be made with the lays.

Impression systems – reel-fed

In order to print gravure successfully on a web-fed machine the impression unit must be sufficiently robust to withstand the pressures exerted and to maintain these pressures when the machine is running at high speed. With many of the light duty and smaller width machines the impression units are hand operated, being raised and lowered by means of worm threads or eccentric cams. Larger machines have push-button controlled pressure systems which may be mechanically, pneumatically or hydraulically applied. Initially there are two principles of applying pressure: 1, rigid impression; and 2, floating impression.

The rigid impression unit is a very positive method of applying pressure and once applied has no resilience in its mechanism. Only a displacement of the rubber covering on the impression roller will cater for a web join passing through the nip and in the event of a web break, several thicknesses of paper may wrap around the roller to cause a jam or a burst roller.

The floating impression unit is lowered until a kiss contact with the image cylinder is made. This point is zero loading and there is an indicator which can be set when the correct pressure has been established. Additional pressure is applied, depending on the substrate to be printed. Springs in the impression block compress to apply the extra load and in the event of joins passing through the nip or a wrap-round taking place, the give in this system will reduce the possibility of burst rollers or sheared gears. Pneumatics or hydraulics can also be used to provide a system that can be accurately measured, set and constantly repeated. These systems are often self-compensating and will align themselves with a cylinder to exert an even pressure along the whole of its length.

On smaller, light-duty machines the pressure unit comprises a rubber roller which is lowered in order to print and raised to allow the machine to idle. The distance the pressure unit travels must take into account the minimum and maximum cylinder diameters and adequate clearance must be allowed between the image cylinder and the web to avoid inking the web when the doctor is lifted. Some units are raised and lowered by motor, with cut-out mechanisms for preventing an overrun and final settings made manually. Most press designers employ the principle of moving the pressure unit in relation to the printing cylinder, but an alternative method is to shift the cylinder complete with doctor and inking system, which has the advantage of maintaining web tension when the pressure is released. Many pressure units on smaller machines have no means of indicating the position of the unit for re-location or the amount of pressure being applied.

When greater pressures are required the rubber roller may be reinforced with a steel back-up or 'bowl' roller. The bearings and shafts must be substantially constructed to withstand the forces exerted upon them and to avoid bowing of the cylinders, which would result in pressure loss in the centre of the unit. Bowl rollers are used extensively on magazine machines and normally provide adequate pressure for the range of papers printed, but with wider cylinders deflection becomes more likely. To minimise deflection some press designers offset the bowl and impression rollers instead of aligning them in the traditional top dead centre position. On packaging presses, which are generally narrower than magazine machines, two back-up rollers may be used to give more stability, but their use must be weighed against the additional heat created by the extra roller. Turret systems are often used on web-fed proving presses where web widths are changed frequently.

There must be a gap between the back-up roller and the impression roller in the raised position, as the rubber swells with long periods of running and will develop a flat if left in contact while standing. In some machines the impression roller has a hollow centre connected to a water supply to keep it cool while running. On multi-unit machines rollers may all be of identical size, which makes them conveniently interchangeable. Alternatively they may be progressively stepped up in diameters by increments in the order of 0·3 mm for the same reasons that image cylinders are stepped up, as explained in chapter 2.

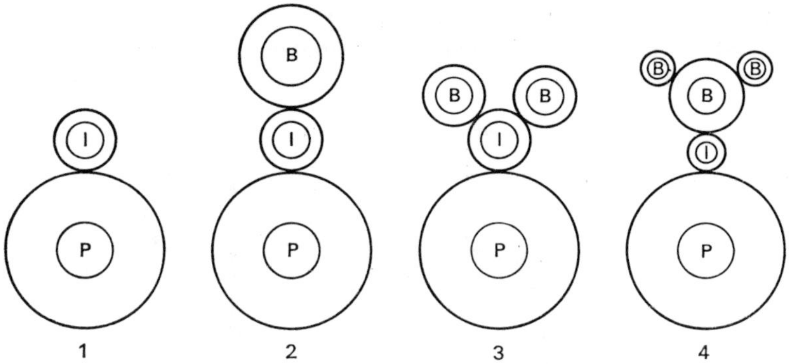

4.8 *Gravure impression systems. (1) Single impression rubber roller. (2) Impression roller with back bowl roller. (3) Impression roller with two backing rollers. (4) Turret system. I, impression cylinder. P, printing cylinder. B, bowl or backing rollers.*

Impression roller setting. Before locating the roller in the machine it should be checked with calipers or micrometer gauges, and once in the machine any system that will ensure accurate contact to the printing cylinder may be used to set pressure. Incorrect setting will affect print quality and register, while excess pressure may alter the shape of the roller, generate excessive heat and cause a burst. On wider machines, mechanical pressure units are constructed to enable one side to be disengaged and the other side adjusted to align the impression roller with the image cylinder. Modern machines incorporate calibrated dials which assist in accurate setting and portable pressure-sensitive gauges can be used to make several checks along the nip. When applying pressure for running it is advisable to start at the reel end and work progressively along the machine, ensuring that there is no slack paper between the units.

Offset gravure

This method has been developed for use on flexographic machines, where a continuous rubber roller of about 40° shore is mounted in place of the plate cylinder and ink transferred to it from a gravure cylinder. Circumferences are usually kept below 700 mm to avoid ink drying before transfer to the substrate. Conventional inking, doctoring and drying systems can be employed and press speeds from 2–3 metres per second obtained, depending on the substrate and drying rate of the ink. It is essential that all the ink be removed at each revolution to avoid build up on the blanket, and highly pigmented inks are used, well retarded, to ensure complete transfer. Colour-to-colour register is usually a manual operation, employing screw-thread adjustment for lateral movement and planetary, differential gears within the cylinders for circumferential movement.

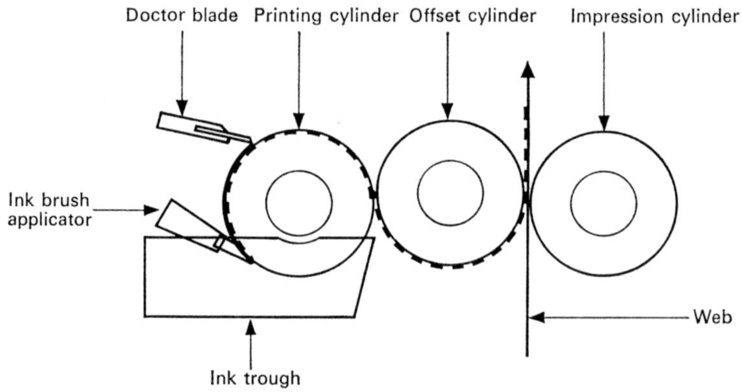

4.9 *Arrangement of printing unit for offset gravure. The image cylinder is shown transferring ink on to a rubber blanket, which then deposits the ink on to the substrate.*

5. Makeready and running

The sheet feeder is the part of the machine which most affects the running speed. For this reason its efficiency and suitability for the job in hand must be most seriously considered. If a feeder is rated at 8 000 sheets per hour but only conveys 6000 then this is immediately a 20 per cent loss of production without considering any other factors.

Manual feeding of machines is now virtually obsolete, but it is worth noting the procedure for hand feeding because, like most other automatic devices, the mechanical feeder is only designed to perform the same operations, but faster and more accurately.

PRINCIPLES OF SHEET FEEDING

The human feeder uses his two hands independently. The right hand separates the top sheet from the pile with the forefinger and thumb and passes it to the left hand, which positions the sheet to the two front lays, then pushes or pulls the sheet to the appropriate sidelay. A delicate touch is needed to ensure that the edges of the sheet meet all three lays without buckling the paper.

The human feeder is also responsible for detecting torn or damaged sheets and for tripping or stopping the machine when the pile is finished, or if for any reason whatsoever a good lay cannot be made.

For this reason all the machine controls for starting, stopping, and tripping were located close to the feeding position, often employing foot controls to keep the operator's hands free for feeding. The influence of hand feeding is still evident in most sheet-fed machines by the choice of operating side and the location of machine controls at the centre of the operating side, even though the control emphasis has passed from the side of machine to the ends.

Automatic sheet feeders

The practical limit for hand feeding is around 2 000 sheets per hour, and actual rates are usually considerably lower, hence the need for mechanical devices. As with hand feeding there are three main control areas; sheet separation, feedboard, and machine control.

Sheet separation may be accomplished by frictional fanning-out of the sheets in the pile, by pneumatic control, or a combination of these two methods. On a small machine the separator may be an integral part of the machine, but for larger sizes of sheet, where more sophisticated equipment is required, the mechanism may be made by a specialist firm working in collaboration with the press manufacturer.

Friction separation is now largely out of favour due to its limitations in respect of speed and accuracy, more complicated setting-up and risk of marking the stock. No new models have been designed for several years and this system must now be considered obsolescent except for a very few limited applications.

Pneumatic separation requires an air supply either in the form of a simple air pump operated from the machine's own mechanism, or a separate compressor driven independently. A combination of blown air and suction is used in every instance. Low pressure air blast is directed at the top few sheets in the feeder pile, and its effect is to separate these higher sheets with a layer of air between each. Suction nozzles then lift the top sheet from the pile and forward it to the machine over this air cushion, where the feedboard controls take over. The air pressures required are usually in the order of 50 kN/m^2 and some adjustment is usually desirable to allow for difference in stock from thin papers to heavy boards.

Front separation. The separator mechanisms may be located at the front of the feeder pile for smaller sizes of sheet. The suction mechanism can then comprise a row of suckers on a common air pipe, mechanically controlled to drop

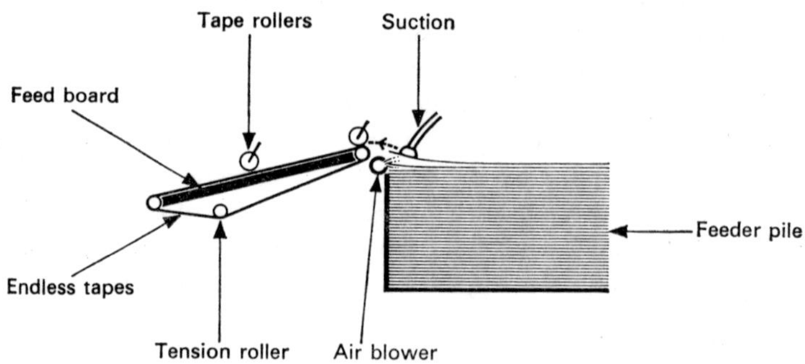

5.1 *Principles of front separation feeders. The air blower is positioned to separate the top sheets in the pile. The lifting suckers take the top sheet and forward it to the feedboard. Tape rollers and endless tapes take over sheet transport on the feedboard and the suckers can lift the next sheet as soon as the first has passed.*

down towards the sheet as it is separated by air blast. The sucker bar then forwards the captured sheet to the feedboard. A tilting action may be used while forwarding the sheet to ensure that it is peeled cleanly from the pile, and additional aids such as small springs are used which allow the top sheet only to pass, while retaining the lower sheets.

Back separation. If larger sheets were fed by the front separation method they would need to travel proportionately faster to maintain the same feeding rate; *eg* an A0 sheet would have to be accelerated to an average velocity of 1 700 mm per second to achieve a feeding rate of 7000 sheets per hour, whereas a feeder designed for a maximum size of A3 would only need to achieve 600 mm per second.

For this reason, on larger machines the separating mechanisms are situated at the back of the pile. In this way a second sheet can be separated before the first has cleared the pile, and the sheets travel to the feedboard in a continuous, underlapped stream. Faster separation can be carried out at slower paper speeds than with front separation feeders, but of course the mechanisms are more complicated.

Two sets of suckers are generally used on back separation instead of the sucker bar on the front models. One set will lift the top sheet once the air blast has made the initial separation and the second set takes over to forward the sheet to the feedboard. Air blast must also be stronger, not only because the sheets are larger but because the air must blow right through to the front edge of the sheet to ensure that it clears the pile and travels smoothly on its cushion of air to the feedboard. Separator springs can also be included and the sheets held in place by a foot which drops upon the pile as soon as the top sheet is separated. One separating head should be sufficient for most purposes, but extra-large machines sometimes require a second head.

Another essential feature of any mechanical feeder is the pile height control. This is the device which automatically raises the height of the stack as the top sheets are used. A sensing device is arranged to touch the top of the stack and operate the hoist mechanism when the pile drops to a critical height. A motorised hoist will be a real advantage with deep piles, and a pre-loading device will allow continuous running. A really efficient pre-loader should enable a fresh pile to be built up while the machine is running, so that it eventually supports the original stack of paper, enabling the machine to be run without any stopping at all.

Feedboard control takes over as the top sheet is presented by the forwarding suckers. The aim is to take the sheets smoothly and precisely to the front and side lays. For forward travel most systems still employ endless straps running down the feedboard with a set of free-running wheels riding upon them. This method, though comparatively simple in principle, requires some time and skill on the part of the machine printer when changing size. It is certainly an area with scope for development. A few manufacturers have devised more positive control systems by means of travelling grippers. These are quite efficient for single-sheet feeders, but more difficult to apply to the larger sizes where the sheets must travel along the feed-board in an underlapped stream.

Careful presentation to the front lays is essential to avoid buckling the front edge of the sheet. Here again the slower travel of the stream feeder is an advantage. Some manufacturers have even incorporated a variable-speed action to the feeding system, so that the sheets are laid very gently at the minimum possible speed.

Having arrived at the front lays, side register must be established. This can be done with small sheets by a moving sidelay which pushes the sheets across the board to a set position, but to avoid buckling on large sheets they are pulled across the board by wheels or fingers to a stationary sidelay. When the sheet is laying flat on the feedboard and touching the three lays it is ready to be taken by the machine and the feeder has done its job.

Machine control. The automatic paper feeder is expected to take over many of the ancillary duties of the human feeder. These include tripping the printing

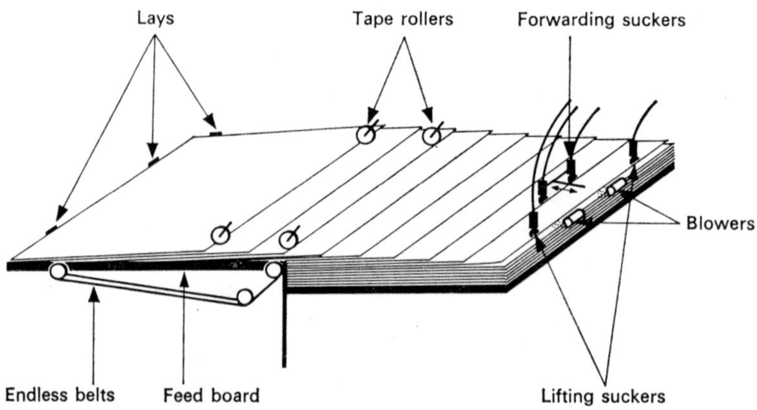

5.2 *Principles of back separation feeders. The blowers separate the top sheets in the pile. Lifting suckers complete the separation of the top sheet from the others. Forwarding suckers take the top sheet, on its cushion of air to the tape rollers, where it comes under feedboard control. As soon as the back edge of the top sheet is clear the lifting suckers can pick up the next sheet.*

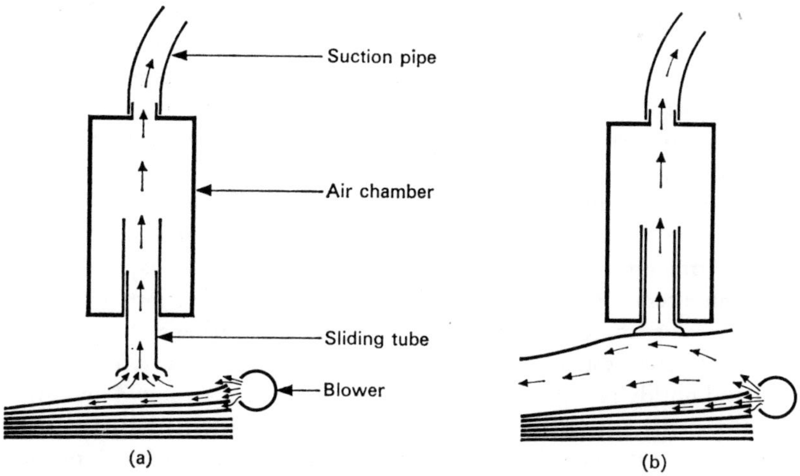

5.3 *Action of lifting suckers (detail of 5.2). (a) The arrows indicate direction of air movement. Air passes freely through the air chamber while the entrance to the sliding tube is free from obstruction. (b) When the top sheet contacts the entrance to the tube a vacuum is formed. The tube telescopes into the chamber, taking the sheet with it.*

unit when no sheets are being fed (missed-sheet detector); tripping in the event of misregister (front lay and sidelay checks); avoiding the feeding of more than one sheet at a time (two-sheet trip); preventing buckled sheets from passing through the machine (crumpled sheet check); and stopping the machine if any sheets pass over the lays (overshoot stop).

These detecting devices may be incorporated in the machine in a variety of ways, usually based on fairly elementary mechanical, pneumatic or electrical principles. A two-sheet detector, for instance, can be an adjustable gate, controlled by a screw-thread to allow one sheet to pass but not two. Front-lay registration can be checked by a photoelectric cell device which records the arrival or non-arrival of the sheet at the lays.

Having detected the fault, the device must be linked to the machine controls to take the appropriate action such as bringing the printing unit off impression to avoid a 'miss' on the impression cylinder, lifting the rollers from the forme, stopping the ink flow, cutting out the paper feed and possibly operating the machine brake.

It is not always essential, or even desirable, to stop the whole machine in the event of a fault. For instance, if a missed sheet occurs on a multicolour machine there will be several sheets at various stages in the machine and an immediate stop would probably waste all these sheets. Here a better solution is to trip only the feeder, allowing the machine to process the sheets still passing through, while the printing units are tripped in sequence. A really efficient control system related to paper feed is essential on any machine to save time, avoid paper and ink wastage and to allow the printer to concentrate on print quality.

Sheet feed and transport includes all settings from the feeder through to delivery. Gripper settings should be checked for change in stock caliper. Spring settings may be self-accommodating while positive adjustments may require an alteration. On one multicolour press all transport grippers are linked so that a single adjustment accomplishes a re-setting right through the machine transfer system. Machine delivery should be set carefully to avoid set-off and untidy piling which damages the stack edges. Slow down gripper systems and the use of controlled air eddy currents round the delivery help to stack work satisfactorily on large presses. Final setting of such devices is not possible until the machine is on the run at speed, ink and moisture alter the stacking characteristics.

LETTERPRESS MAKEREADY

In its wider sense makeready is the whole process of preparing a job up to the moment of running. The printer is concerned with three main variables: printing surface, ink and the paper or other substrate. The machine is a catalyst whereby these items are brought together to produce print, and it must be correctly programmed for each different end-product.

As with any exercise requiring technical skills, the printer's first duty is to understand the precise nature of the end-product required. Comprehensive instructions are therefore essential, and as much information as necessary must be obtained before the work is started. Attention to this aspect will avoid delay, costly mistakes and above all will ensure that the finished product is suitable for its intended purpose. The following procedure outlines the makeready cycle.

The printing machine must have critical attention when preparing a job. It should first be checked for cleanliness, correct lubrication and mechanical condition. Then the printer should carry out a systematic setting-up of the main sections – printing cylinders, inking system and paper control. The impressing surface must be carefully dressed and checked for condition. The printing surface carrier must be thoroughly cleaned to avoid the possibility of trapping foreign matter below the forme. The inking rollers must be finely set to distribute the ink and lay it evenly on the printing surface. The paper control system should be set to the size and caliper of the printing material, bearing in mind any special characteristics such as coated surface or porosity.

The printing stock can now be loaded into the feeder and a number of sheets run through the machine to check the accuracy of the settings. This procedure also provides an opportunity to detect paper faults such as curl, wavy edges and static electricity.

The printing surface may then be prepared. A flat forme must be checked for caliper, loaded into the machine and planed down on the bed to ensure accurate contact and absence of springing. Secure locking up without over-tightening is essential. A check should be made to ensure correct relationship of forme to paper. Rotary plates must also be checked for caliper and prepared as described in Chapter 2, before mounting on the plate cylinder. In the case of one-piece wrapround plates or plate-sets on carrier sheets, mounting and positioning will be the work of a few minutes, but where a number of individual plates are to be mounted this can be a lengthy operation. Having located the plates as accurately as possible, a final check is made to ensure conformity to cylinder.

Ink must be checked for colour and quality before loading into the duct. The duct roller should be rotated to check and adjust the amount of ink being metered to the distributors, and a measure of ink fed to the rest of the system. An impression can then be taken.

The first pull from the machine will reveal the degree of accuracy of the printer's settings. It is unlikely that this first print will be completely satisfactory. In fact it may be considered good practice to run a number of waste sheets to allow the settings, particularly inking, to settle down. From the first clean print however, the printer will be able to determine the adjustments necessary to each of his three variables in order to obtain a satisfactory print. His first pull should also be sufficient for some initial checks to be made by the reader and quality control.

The time taken to make the appropriate adjustments to impression, position and inking controls will vary considerably according to the nature of the work. For instance, pressure equalisation may be necessary on illustration work but not on bookwork, colour control will be more subtle when overprinting transparent coloured inks, and colour halftone work will demand meticulous registration compared to single-colour jobs.

Pre-makeready

Makeready time will be substantially reduced by a properly organised pre-press section. Of the three basic functions affecting print quality – pressure, position and ink, only inking need be entirely a machine setting. The other two can be positively established before the printing surface is brought to the machine. It has been shown that pressure is essentially a question of caliper, and register is of course a matter of calibration.

The function of the pre-press section is to provide printing surfaces which have been accurately prepared for caliper and position. Seen in this light, pre-press work is initially a matter of measurement, and again the importance of adequate instructions must be stressed. There are two essential aspects to this work, one is to measure and the other is to prove.

The equipment for pre-press work comes under three basic categories:

1 Positioning devices. There are two basic types: (a) calibrated systems for locating plates and type formes for monochrome or first-colour workings; and (b) register comparison devices for aligning other workings to the key. Positioning is discussed more fully in chapter 2.

2 Caliper equipment. These are micrometers and special-purpose gauges for the precision measurement of type, blocks and plates. To these must be added pressure adjusting equipment for the preparation of overlays and interlays where required.

3 Proving machinery. This is essential for obtaining proofs of the prepared work so that the accuracy of the preparation can be confirmed. For flat formes the roll-over type of press is used where the cylinder travels along the bed pushing the inking system in front. Rotary systems may be designed for proving only, or they may also incorporate a mounting system for individual plates to be mounted on a removable machine cylinder or a carrier sheet.

It must be obvious that the pre-press section is really a part of the quality control system in a printing house, and the scope of the section could extend well beyond the basic functions mentioned above. Other aspects may include materials testing, colour mixing and matching, inter-departmental co-ordination, supplying proofs to customers and supplying key sheets to the machine room.

Register work

The essential difference between colour halftone printing and monochrome work has been emphasised in previous chapters. Technicians engaged in any aspect of close-register work must understand the implications of working to tolerances of hundredths of a millimetre. The concept of plate register has been explained in Chapter 2 and it is essential to have a set of printing images prepared with sufficient accuracy for register to be obtained.

Granted this basic requirement, we must then consider: 1, the machine aspects; 2, the effects of ink and paper on register; and 3, printing technique.

Machines

Machines must initially be built within the specified tolerances, and when in motion they should be free from vibration or other false movement. Impression mechanisms should not distort measurably under the printing load and gears and bearings should be free from any play or false movement while the impression is being made. The slightest tendency to shudder in a machine part before or during impression will mean that results cannot be exactly reproducible.

Settings on the machine must be finely calibrated so that minor adjustments to paper position and printing plates can be readily made. Ideally it should be possible to make all adjustments while the machine is running to avoid time-wasting stops and to allow for slight variations which may occur at different machine speeds.

The paper control system should be particularly well engineered to ensure that each sheet is presented to precisely the same position at every impression. Paper must be handled firmly to ensure accurate registration, but not roughly to cause damage, creasing or buckling.

Sheets are generally pulled through the machine by sets of grippers. These must hold the sheets firmly, but without marking. Spring loading of each gripper is advisable with an overall tension or caliper adjustment to allow for differences in stock caliper and substance. A satisfactory setting for a thin, hard paper like bank may cause marking on a thick soft board such as pulpboard. Impression cylinder grippers need particular attention because they must pull the sheet through the printing nip without slippage, yet the space they occupy on the front edge of the sheet must be limited to keep the non-printing area to a minimum.

The travel of the sheet through the printing nip must be free from differential stresses. If the cylinders are not both made up to the pitch circle of the gears, the paper may slip under impression and slur will result. The printing stock must be considered as part of the impression cylinder for this purpose, and allowance for its caliper made when packing the cylinder.

Slur is more likely to occur when printing fairly full images such as half-tones, where the printer overpacks one cylinder or the other to obtain a firm impression. Where this need arises the extra pressure should be obtained by over-packing both cylinders equally so that their peripheral speeds remain equal. On a flatbed cylinder machine, slur due to overpacking may be avoided by applying additional packing to the forme only. The lateral speed of the forme will not be altered and must still match a correctly-dressed cylinder rotation.

Ink and paper

Ink and paper (or other substrate) can greatly influence the efficiency of register. A tacky ink will pull at the printing material, affecting its travel through the machine. Variations in ink tack can be due to changing machine speed, different batches and colours of ink, and temperature changes. On multi-colour machines particularly, tack control is one of the most critical aspects to be understood.

Paper is notoriously hygroscopic and will change its dimensions according to the moisture available in the atmosphere. Some papers are more sensitive to atmospheric changes than others. Permanent physical distortion can also occur with soft-furnish papers under heavy impression. Evidently quality control must be more thorough for fine register work and atmospheric control must be practised to ensure paper stability.

Technique

Technique employed by the printer is a critical factor in fine register work. Meticulous attention to detail in setting up the machine and preparation of materials is vital. To contain register variations to less than 0·1 mm over the length of an A0 sheet requires an accuracy of better than 0·01 per cent. When preparing the cylinders of the printing units, a caliper variation of 0·15 mm in the plates or cylinder dressing will change the overall cylinder circumference by about 1 mm which is at least ten times the tolerance permitted.

Colour halftone printing

Accurate register is one aspect of good quality colour halftone printing. The other two are accurate colour rendering and freedom from printing faults.

Colour control starts with the original artwork which must be faithfully reproduced as corrected separations in the process studio, translated into dot formations of equivalent value by the platemaker and the resulting relief plates covered with a controlled film of ink by the printer, so that when transferred to the substrate it presents the same appearance to the eye as the original copy. The visual appearance of the print is determined by 1, the optical properties of the ink ingredients; 2, the thickness of the ink film; 3, the substrate; and 4, the viewing conditions.

Process inks must be formulated with pigments which are not affected by other ingredients, strong light, the substrate, the printing surface or any other materials with which the ink may come into contact, such as varnish or laminates. The spectral reflectance of different sets of process inks varies considerably, and none are perfect. For this reason rigid standardisation is necessary at all stages of production. This is not only to ensure fidelity to the original, but so that correction factors can be introduced into the platemaking processes whereby deficiencies in the colours are equated. It is a fact that uncorrected plates will never produce faithful reproductions of the original, due to the imperfect spectral ranges of the process inks. If the platemaker is aware of the exact spectral distributions of the inks to be used he will be better able to produce a set of plates which correctly interpret the copy.

The machine printer will usually work to a proof of the job supplied by the platemaker. This will be in the form of a set of progressive prints showing the ink densities required at each stage to produce the desired result. Control devices should be included in the progressive proofs such as colour bars comprising solid patches and proportional screened sections. These can be used for visual reference and in conjunction with densitometer readings to ensure correct ink film thickness. The progressives should also give relevant information regarding inks used, order of printing, screen values and angles, and viewing conditions.

The substrate is a major influence in colour halftone printing. It should be remembered that the multicolour effects of process work are obtained by progressive subtraction from the full spectrum as represented by the white base. For this reason the nearness to perfect whiteness of the printing stock is a measure of its quality for colour work. The lighter tones of a colour print are seen as an optical mixing of coloured dots with the whiteness of the paper. The farther from white a paper is, the more degraded will be the final result; therefore any stock with a distinct colour of its own will not be capable of reproducing an accurate colour print.

Other properties of the material will also affect the appearance of the printed result. These include smoothness, gloss, opacity and surface texture.

Where colour halftone printing must be done on a non-white substrate it is common practice to lay down a solid white layer first, which then becomes the base for the colour print. The white base may also serve to overcome other deficiencies of the substrate; for instance, acting as a filler on rough surfaces, or as a key where adhesion is a problem.

Viewing conditions will largely control the appearance of the print under examination. Apart from a sufficiency of light, the quality of the light falling on the print must match that illuminating the original when it is approved for colour value. Agreement on the illuminating medium should be made before proceeding with any work where faithful colour reproduction is desired. This will help to settle differences of opinion as the work progresses, and will counter the effect of metamerism, where two colours appear to be identical when viewed by one light source but quite different under another illuminant.

Readers are referred to British Standards BS1480 (1949), BS2650 (1955), BS3020 (1959), BS4160 (1967) and BS4666 (1971) which all refer to process colour ink standards. BS950 (1966) relates to viewing conditions in the graphic arts, and makes recommendations for standardising conditions. The ultimate illuminant of the printed work should be borne in mind when making a qualitative assessment. For instance, most mail order catalogues will be looked at by customers in their own homes under tungsten lamp lighting, while posters for display in the underground railway will be seen by fluorescent lighting.

Wet-on-wet printing

Multicolour machines are particularly susceptible to small changes in ink-film thickness as the thinnest practicable layers must be run to keep the weight of ink on the paper to a minimum. In four-colour work (three colours and black) it should never be necessary to lay down more than one 100 per cent solid colour. Four superimposed films should never exceed 270 per cent of solid colour if trapping, drying and marking-up problems are to be avoided.

To this end the platemaker must practice undercolour removal where his plates are to be used for wet-on-wet work. Wherever a solid black is printed the corresponding areas in the colour plates are superfluous, and the image may be reduced proportionately. This practice will not only reduce over-printing problems and improve print quality, but it can also save a considerable amount of ink on a long run and some makeready time.

Successive ink films must be laid down before the previous layer has dried hard. This trapping procedure requires careful tack adjustment for wet-on-wet printing so that the maximum amount of ink is transferred from the printing surface to the previous ink layer on the substrate. Special ink-sets with progressively reduced tacks can be provided by the ink-maker, or adjustments may be made to constant-tack inks by the machine printer. Assessment of tack may be made by empirical comparison between colours in a set, or by reference to a set of tack-graded pastes which are supplied by some ink manufacturers. More specific measurement may be made on a tackmeter which measures the pull of the ink under specific conditions at a given roller surface speed. The trapping value of an ink layer will also vary with its thickness. This is a further reason for strict control of ink film thickness by ink monitors or densitometer readings.

Colour sequence

The practice of undercolour removal, tack control and use of correct screen angles will enable the printer to choose his colour sequence to suit the job in hand. Traditionally the lightest colours are printed first so that discolouration of following colours is kept to a minimum in the event of imperfect trappings. However, as yellow is the most difficult colour to assess visually for colour value and printing faults, it is often considered an advantage to be able to vary the conventional yellow, magenta, cyan, black sequence.

Moiré is the name given to an unwanted pattern appearing in process work due to the visual clashing of two adjacent halftone screens. Correct choice of screen angles usually avoids this unpleasing effect but it may occur in midtone areas when superimposing wet ink films.

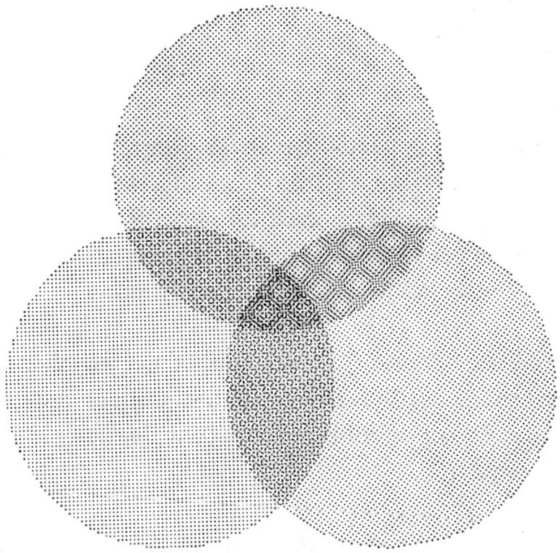

5.4 *Moiré patterning. This is a visual effect caused by the super-imposition of screens at different angles. Subtle differences in angle can accentuate this effect, therefore careful attention must be paid to selection of screen angles for colour halftone work if unwanted patterns are to be avoided.*

When printing process work on two-colour or multicolour machines it is often advisable to keep the colours in pairs with the screen angles as far apart as possible. With the common system of yellow at 0°, cyan at 15°, black at 45°, and magenta at 75°, this means keeping yellow and black as one pair and magenta and cyan as the other. This sequence has the added advantage of keeping the dominant red/blue colours together to minimise the possibility of misregister there.

Progressive proofs for wet-on-wet printing must be produced in the same order that the printer is to run them. As the speed of printing can be critical to colour reproduction, the prover should try to match production conditions to the best of his ability.

Letterpress printing faults

Most faults in printed matter can be detected before running on. Two functions of a quality control section should be to devise means of avoiding faults, and to detect those which do occur at the earliest possible moment. This approach will reduce to a minimum a third duty of finding faults after the work has been done.

Three groupings of faults are dealt with here; 1, those associated mainly with the machines; 2, printability faults – those concerned with ink/paper/plate relationships; and 3, post-printing faults.

Machine faults

The fundamental defects of over- or under-impression, misregister and incorrect colour may all be present in the first pull from the machine, but should be corrected by the systematic procedure of makeready. There should be no excuse for these faults being present at the beginning of a run, but they must be looked for frequently while the job is progressing.

Over impression may be seen as indentation of the stock by metal plates or distortion of the image with resilient plates. Causes may be movement of cylinder dressing, swelling of rubber plates by solvent or ink, or partial lifting of poorly-secured plates.

Under impression will be seen as weak areas on the print (not to be confused with poor inking), or a general lack of firmness in the image and may be caused by plates or packing settling down after many impressions, permanent compression if the elastic limit of the packing is exceeded, or general wear of either surface.

Misregister can be due to plates shifting, movement in machine settings, ink tack variations, paper instability or static electricity. The printer should prepare for the possibility of misregister by printing a number of sheets at the beginning of the run when the position is established and feeding them again through the machine at regular intervals to check for register variations.

Colour changes can be detected by reference to a master sheet kept by the delivery for this purpose. Visual inspection is good enough for most purposes, but where minor changes in hue are critical the colour strength can be confirmed by the use of a reflection densitometer. Colour variations may be caused by changes in ink film thickness due to temperature or speed changes affecting viscosity and tack. Carryover of ink in wet-on-wet printing will also affect colour values.

Slur has been referred to as the result of incorrect packing of plates or impression cylinders. This can be recognised as a distortion of the printed image usually towards the back but sometimes in the opposite direction. Dots in halftone printing may appear to have tails at the back edge, indicating overpacking. This fault will not only mar the appearance of the print but will cause misregister and premature wear of the plates and cylinder dressing. False movement of plates, packing or cylinders will also cause slur.

Wipe is recognised as a ridge of ink at the image edge, often combined with a lack of colour on the printing surface itself. This is caused by a false movement between the forme and inking rollers and may be the result of incorrect setting, eccentric rollers, skidding, lack of drive to the rollers, loose bearings or swelling during the run due to temperature changes, friction or other influences.

Repeat is a fault of ink distribution, recognisable as a fainter image in some parts of the print where the forme rollers have not been recharged with ink after completing a full revolution on the forme. This is an inherent design fault in some machines, but may be aggravated by poor roller setting and thin-bodied or transparent inks.

Ghosting can be confused with the previous fault, as a repeat will often appear as a lighter image of one part of the forme in another area. Strictly, ghosting is an effect due to a printed image on one side of a sheet affecting the drying of the print on the other side. This shows up as a 'ghost' of the first image. Ghosting is also the term used when the cylinder dressing does not recover in time for the next impression, resulting in a weakened image.

Misting occurs when an ink film is split between two outward-turning rollers. The film is first pulled into fine filaments between the rollers and upon rupturing the strands throw off fine droplets of ink. These droplets often have an electrical charge which prevents them from returning readily to the rollers and an accumulation of them causes a haze in the vicinity of the machine. This is more likely to occur with viscous inks, fast-running machinery and thicker ink films. The mist may be contained by enclosing the inking system, or extracted and recovered by a suction/filter system. Other methods are to blow the ink back on to the rollers by air jets positioned at the outward nips, or to neutralise the charges on the droplets with ionising devices similar to those used to cure static electricity in paper.

Set-off is the transfer of ink from the print to another surface. This is a particular nuisance when delivering sheets in stacks where a wet print may transfer ink to the sheet above it and so on throughout the pile. Quick-setting inks have reduced this problem considerably, but an anti-set-off device is still standard equipment on many machines. Anti-set-off devices spray a finely divided powder on to each print as it reaches the delivery so that the sheets do not actually touch each other.

Second-impression set-off occurs on perfectors, where the paper is backed up immediately after the first side is printed. Where it is likely to occur, ink films should be kept as thin as practicable and the second impression cylinder covered with a blanket having a non-ink-receptive surface. Suitable surfaces are non-receptive plastics such as nylon, silicones or even a coating of tiny glass beads as produced by one blanket manufacturer. Anti-set-off sprays would be a nuisance here due to the build-up of powder, but a scavenging roller is sometimes effective on slow-running machines, wiping the cylinder with a thin, cleaning fluid.

Guttering is a fault of badly-set bearers or overpacked formes. The printing surface is thus made to take the whole impressing force, with the cylinder dropping into the margins between forme elements. Hard edges on the printing image are the first sign of this fault, and the cylinder can be heard to rumble across the forme. Premature wear of the forme, packing and machine are inevitable if this fault is allowed to develop.

Printability faults

These are the result of incompatability between the printing variables. They can nearly always be avoided by employing fairly simple materials-testing practices.

Filling-in is most noticeable on halftone work where ink accumulates between the dots until the screen is completely lost. This common fault can be caused by any one of a number of factors, or a combination. Too thin an ink will run down between the dots; fluff or surface dust may clog a tacky ink and fill in; or poorly-ground pigments may be the cause. The plates may be at fault if they are not etched deeply enough, or the screen value may be too fine for a rough-surfaced paper. Other possible causes are ink drying too quickly or anti-set-off powder getting directly or indirectly on the plates. The fault will be aggravated by badly-set rollers, excess impression and the conditions which cause slur. The printer's obvious remedy for filling-in is to stop the machine and scrub out the forme, but in bad cases this can be extremely time-wasting. The correct procedure is to diagnose the cause and take the appropriate corrective action.

Fluffing is particularly noticeable in solid areas where accumulations of paper fibres will mar the smooth appearance of the print and may cause filling-in of halftones. Tacky inks are generally responsible, in conjunction with loose-surfaced stocks.

Picking is a fault of the same nature as fluffing but occurs with coated stock, being a result of tacky ink pulling the coating from the surface of the paper. With both of the above faults, if the stock cannot be changed the ink must have its tack modified by the addition of a suitable reducer.

Mottle is seen as a speckled appearance, particularly unsightly in solid areas. This uneven effect may be due to selective absorption of ink by an inconsistent paper. Soft inks are liable to mottle as their lack of cohesion allows them to be squashed into tiny 'puddles' under impression, from which they do not recover. This is particularly evident with thicker ink films, transparent colours and platen impression.

Reticulation or crawling is the result of complete incompatability between the ink and its substrate. This is a rather dramatic fault in which the ink gathers itself into small droplets on the surface of the material instead of forming a film. This can only occur if a complete error of judgment has been made such as using a water based ink on a plastic film.

Refusal has some of the characteristics of reticulation, but occurs when a superimposed ink film will not adhere to the lower stratum due to the previous layer having dried hard. Minor cases may appear mottled, while severe instances may result in reticulation.

Show-through is an optical property of the paper, related directly to its opacity. It is the degree to which the print on one side of the sheet may be seen from the other side. The nature of the work and its intended end use will determine the significance of show-through. This property can be measured by comparison with a set of graded papers of known opacity, or by use of an opacimeter head on a densitometer.

Strike-through is often confused with show-through, but is in fact the appearance of ink on the reverse side of the sheet due to an over-thin ink or an extra-porous paper. This defect can occur even with an opaque paper, but where both show-through and strike-through are present the effect will obviously be particularly displeasing.

Post-printing faults

Blocking is really an extreme form of set-off in which the sheets stick together due to ink drying in the stack. This fault also occurs with rewound reels, where the efficiency of drying cannot be checked by taking out a copy. A thorough understanding of the drying mechanism and its control is essential to avoid this problem. A change in one variable, such as machine speed, can make the difference between blocking and trouble-free rewinding.

Haloing shows up as a tint around the image boundaries due to the migration of vehicle through the stock after printing. This can sometimes occur after a considerable time has passed and may completely change the appearance of a print, particularly halftones. Unsuitable inks on an absorbent paper may give rise to this fault, or an over-reduced volatile ink in which the solvent phase has not been thoroughly evaporated.

Bleed is a more serious fault of the same nature as halo but in this case the colourant may also migrate after the print has apparently dried. Some dyes are particularly prone to bleeding. This can result in a delayed strike-through or even blocking in severe cases. Tests for bleed must be carried out if the print is to be varnished, laminated, or used for wrapping cosmetics or foodstuffs, particularly soaps and fats which have an affinity for certain ink ingredients.

Powdering occurs when the print has dried. The surface ink becoming loose enough to brush off as a powder. This is caused by disproportionate penetration of the ink vehicle, leaving an unbound pigment on the surface. This can be a very expensive fault to discover if it is not found until the job is finished.

Fading of the print can happen if it is exposed to strong light. Where this is undesirable a lightfast ink must be specified, and the inkmaker will select pigments which do not readily fade. Fading is occasionally caused by exposure to other agents such as gases or chemicals, and where the print is known to be exposed to special influences, tests should be devised to check for any possible ill effects.

Post-printing requirements

Printing is rarely a complete process in its own right, but one of a series of operations necessary to manufacture the end product. In reel-feed production the aim is usually to perform as many finishing operations as possible in-line with printing, but with sheet-fed production very few operations other than printing are undertaken in the machine room. The Print Finishing Department is required to undertake the many and varied procedures which are necessary to complete the work, and it is essential that the work of the two departments be co-ordinated if efficient production is to be achieved and costly mistakes avoided.

In bookwork, for instance, the positions of the pages on the sheets will be determined by the folding procedure in the Finishing Department; sufficient allowance must be made in the margins between the pages for folding and trimming, and the method of binding the book will affect the amount of space in the back margins of the sheets. Each sheet will have identification marks to enable it to be easily assembled and collated after folding.

Labels, showcards, maps, covers, book jackets, cartons and many other types of work may be varnished or laminated and care must be taken to ensure that inks and substrate are compatible with the finishing media. Difficulties could arise from excess powder spray, unevaporated solvents, arbitrary additives to inks, changes in the moisture content of the paper and any other variations from the test piece.

Many wrappers and cartons for food and cosmetics are waxed after printing to give added protection, and similar precautions should be taken as for varnishing and laminating. As most waxes are heated before application and then rapidly cooled, additional checks should be made to ensure that rapid temperature changes do not unduly affect the printed work.

Boxes, cartons, envelopes, point-of-sale material, toys and games need cutting to shape by special dies in the cutting and creasing section. Careful planning is essential to ensure accurate fit between the cutting and creasing die and the printed sheet. It is usually easier to fit the print to the die than vice-versa; the die should therefore be made first and the printed work planned to fit a master taken from the die. Care should be taken to avoid misregister, incorrect paper grain direction and dimensional changes in the substrate.

Work to be embossed must have a sufficiently stout structure to be distorted without rupturing. The ink film must also be strong enough yet flexible to take the contours of the embossed image. Indentation of the substrate by excessive printing pressure may show up as unwanted outlines in the finished article. The register tolerance required will vary according to the characteristics of the finished work.

Bronzing is a process in which a metallic powder is dusted on to an adhesive surface, and is generally considered to produce a better quality finish than printing with metallic inks. The substrate should be tested for chemical contaminants which may cause the metallic powder to discolour, and where overprinting occurs, similar tests should be made with regard to the printing inks. Bronzing is more fully covered in chapter 7.

Whenever printed sheets are to be further processed, the machine printer should mark the positions of his lays on a master sheet to ensure that the sheets are registered to the same positions in subsequent operations. Allowances should be made for wastage due to setting-up and makeready of other machines and processes and a few sample sheets should be forwarded to the Print Finishing Department as early as possible so that any errors may be discovered before running on.

LITHOGRAPHIC MAKEREADY

Early preparation for the run may mean getting customer approval of proofs which are normally prepared in small quantities on machines less costly to run than those used in production. Flatbed offset presses are usually used for this pre-production run, and in their automated forms are frequently used for short run production work as well. Apart from their economic usefulness in handling short runs, they can also be designed to accommodate special materials, ranging from thin plastics to glass, which are not easily printed on rotary machinery. Individual care can be given to each copy for production work, but the preparation of copies for customer approval as targets for work to be machine printed must bear in mind the capabilities of the printing press. The proving method should produce a result comparable to that attainable on the printing machine, using the same inks and substrate which will ultimately be used on the production run. For the same reasons it is advisable that items to be printed together should also be proved together so that the colour balance obtainable on the printing machine is realised on the proof. If more than one job size is to be proved, they should be proved together so that the same colour balance is available as a target. Sometimes large plates are required to accommodate a variety of work and machine proving can be resorted to, usually on machines set aside for this purpose. The advantage of conventional proving is time-saving to the printer by preparing ink mixes and colour sequences. If the colour proof is only intended as a guide to the printer, the retoucher and as a customer's copy, this can be more cheaply prepared by the many photo-development methods available.

All materials should be supplied to the machine. Pre-makeready saves the printer's time and enables many mistakes to be avoided. Scheduling is the main function of pre-makeready and this means that a works progress system can be operated to advantage. Works dockets, plates and instructions should be brought to the machine before the job change, and early paper inspection made to check its condition. A check list supplied to each machine can be used to mark off the receipt of various materials, and the headings of appropriate makeready tasks to be performed. This is useful as a reminder to the printer working alone, but more especially useful on shift work or where a team is employed on larger presses. Works orders are first checked with records for previous performance data, and ink mixes, anti-set-off needs, machine plates, paper and packings inspected and prepared during the previous run. Time and trouble is saved by preparing lay marks, deciding colour sequences to avoid unnecessary changes, checking that materials supplied are correct, and by preparation for any special circumstances arising with the job change.

Lubrication may be on a maintenance schedule but can be checked during makeready. Items removed and replaced during makeready should be lubricated. Washing up and cleaning of cylinders are included under this heading.

Whether ink for the job is matched in a separate department or on the press, a printer is involved. Colour must be matched within the range of stock inks available. The physicist's use of the term 'hue' as confined to pure colours is recommended, coupled with the modifying shading obvious to the eye. Thus redshade cyan is a more exact term than 'warm process blue' and greenshade yellow is more definite than 'cold process yellow'. This does not exclude classification of specific colours with identifying titles such as turquoise, primrose, emerald green etc., but from the point of view of the colour matcher it is useful to consider the primary subtractive hues as points on a colour circle. If the edge of the circle is recognised as colour saturation and the result of mixing any two colours on this perimeter is taken as a point on a straight line between them, it can be seen that the further these two colours are away from each other on the perimeter, the more adulterated their mixture becomes. A mixture of contrasting colours will in theory produce a neutral grey. With this picture in mind, choice of stock inks becomes easier and the required quantity of each can be estimated. Mixing should commence with the colour that predominates in the mix, even if this should be the extender or tinting medium, and assessment should always be made on the stock to be printed.

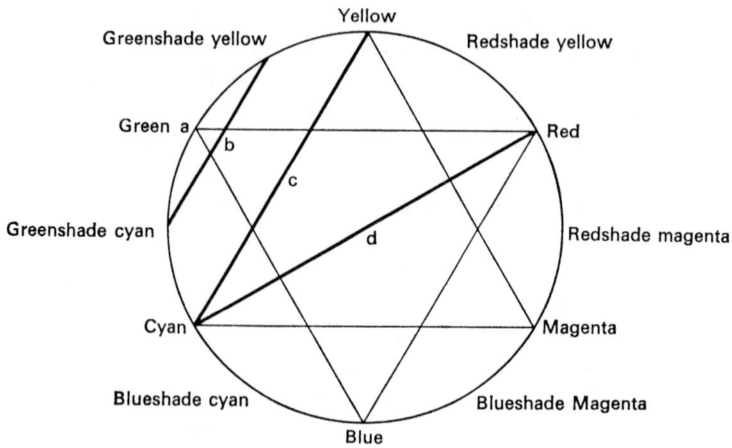

5.5 *Colour mixing circle. White light exposure of copy through the red, green and blue filters produces the cyan, magenta and yellow printers respectively. The latter become the colour printer's subtractive primaries to produce pure hues when overprinted. However, they lose intensity when intermixed, as this colour circle demonstrates. The perimeter represents colour saturation with intensity decreasing towards the centre, so that complementaries cyan/red produce a dark grey mix (d). A mix of subtractive primaries cyan/yellow produce a desaturated green (c). To produce an intense green approaching the pure hue (a) a greenshade yellow and a greenshade cyan must be available (b). It follows that a range of inks outside the trichromatic range are necessary for true colour matching and an understanding of the degree of saturation is necessary to make the right choice when mixing colours.*

Accurate plate fitting is fundamental to work positioning and paper feed, and should be completed before ink is run up in case any faults arise. A sound method of registering the plate is desirable, the simplest system being the employment of a centre gauge for horizontal positioning coupled with the use of scribe marks on the cylinder edge to indicate the first line of print margins. The necessary lining up of the machine plate can be by pre-marking pencilled lines on the gummed plate, or by a punch system which can be arranged to coincide with pin bar systems used in platemaking. Pre-registering systems are also in use and enable most of the work to be prepared off the machine. Fitting the plate square on the machine is important and no attempt should be made to advance one side in comparison to the other without making the necessary compensating lateral adjustment to allow the plate to 'swing'. Failure to recognise the helical nature of this plate movement causes plates to split at the gutters, or pull out of the clamps. Excessive tension should be avoided and short keys or spanners used to avoid overstretching the plate. Plate packings must be checked with a micrometer before fitting, and for multicolour work a sequential reduction of plate packings allows for paper stretch. Modern presses employ lateral and axial external adjustments to be made on the run, and in a few cases helical adjustment is attempted. When the requisite quantity of ink has been matched, the ink duct setting can be made and dampers moistened and fitted into the machine. Colour can then be run up onto the roller train ready for printing at something approaching the correct filmweight.

If all the other tasks are completed first, operator intervention is less likely to cause deterioration of the machine plates. Machine settings are always important in deciding plate wear potential and the less washing out and re-gumming during makeready the better. Adjustment to colour density may be purely subjective, or by density readings on equipment separate from the machine. Normally, however, it is necessary to achieve a supervisor's 'pass' before colour and register targets are set. Registering should be achieved on the plates as much as possible to avoid maladjustment to the feeder. On multi-colour presses it is often essential to register units together, since lay adjustment to the feeder affects all colours printed. Once correct register has been achieved, marks should be put on the first plate to indicate the gripper-edge side of the sheet and at the point of contact with the side lay. These give guidance to the operator who may carry out subsequent processes demanding close register such as folding, cutting and creasing.

Sheet control on offset machines

Paper for litho printing needs sufficient moisture content to give stability and to prevent absorption at the stack edges causing them to expand. Stack temperature and humidity can be checked before use and unless there is pressroom control of these conditions stacks should be covered between printings with protective material such as polyethylene. The moisture content is changed when running through multicolour litho presses and any rapid drying out will be greater at the tops and outsides of the stacks resulting in dimensional variations. Stacks of work should not be aired and restacked between prints to aid drying if register stability is the greater problem. Paper conditioning is recommended where distortion problems occur frequently.

Gripper pads are normally set level across the machine and tension should be adequate to prevent slippage. It should be noted that with spring tension systems, gripper release is earlier with thin stocks and takeover later than with card, and this influences the timing changeovers. For this reason pad height adjustment is often incorporated for various stocks to be accommodated. Circumferential vibration and the consequent tooth chatter causes register error during gripper control and the additional problem of ghosting. Machine designers employ the same techniques to prevent this occurrence as in the avoidance of play and vibration between cylinders, *ie* use of register segments or spring idler gears. Additionally torsional deflection of gripper shafts and smearing of work on cylinder drums is prevented by machine design variations mentioned in chapter 1, and individual adjustment may be made to devices installed for this purpose.

The impression nip is an effective part of sheet transport, and is particularly influential with fragile papers or coated stock. Accurate fitting of offset blankets helps to prevent the paper wandering through the nip – particularly on perfecting machinery of the blanket-to-blanket type. Blanket grain must run round the cylinder with an effective method of squaring used to ensure a 90° tension to the gripper edge. Templates can be prepared for each machine type, and with larger machines gripper and back edges of the blanket are sometimes bowed to prevent slack sides under tension. Slackness at the sides occurs partly because of bowing of blanket bars, and also because of the pattern of mechanical distortion on impression shown in chapter 4. Curving of the blanket clamp edges helps even sheetfeed through the nip, but should not be more than 3 mm for a 600 mm wide blanket, 5 mm on a 1 200 mm wide and 8 mm on a 1 800 mm wide blanket. New blankets should never be marked out from used ones because of accumulated inaccuracy, and marking out must always be from one basic point for the same reason.

Press deliveries require careful setting, especially for large sheets. Mechanical curl, caused by careless handling or through manufacturing problems and reeling, is a nuisance in any delivery. Large areas of solid print alter the structure of one face of the printed sheet and effects a curl, particularly at the back edge of the sheet when the solids lay well back. Additionally the litho printer has to contend with moisture pickup causing curl in the delivery, usually away from the printed side, it may then flatten or curl the opposite way when the paper dries. This condition affects sheet feeding with the next pass through the press. With very difficult stocks, flat laying of high piles is not always possible, and a system of racking may be introduced into the piling arrangements. Where 'cow-catcher' mechanisms are fitted for automatic pile changeovers, racks are effectively used in conjunction with corner blocks or posts to erect a number of small piles in one delivery load so that the full value of a highpile delivery is not lost.

Board does not present the same problems with curl, but its rigidity is more likely to search out protuberances which foul the sheet. Heavy boards retain mechanical curl from their passage round the cylinders and are weakened with subsequent flattening in the pile. Machines for the regular production of carton work should be chosen with this point in mind, and sharp changes in direction of sheet passage avoided. Piles build up quickly and the use of anti-setoff equipment and racking will minimise setoff in the stack.

Metals can be printed on paper printing machines if thin enough, or when laminated to board. Foils for this purpose are usually aluminum having a low modulus of elasticity and consequently prone to curling. Sheets must be kept flat and pressures at a minimum. The natural oxide surface has poor polarity for the reception of lithographic ink and the printing surface is therefore coated with a vinyl or nitro-cellulose material.

Heavy sheet metals such as tinplate and aluminium sheet are printed on machines specially designed for the purpose. The inherent rigidity of the metal requires that it pass straight through the impression nip without curling, although some multicolour machines are designed with large transfer drums between units. Tinplate is basically black in reflectance and requires a white printed or coated base colour for conventional decoration. Poor affinity for ink results from the presence of grease or moisture on the surface, and condensation has to be avoided or the plate will 'frost'. Tinplate must therefore be conditioned before printing to bring its temperature up to room conditions, preferably before stacks are uncovered. Other handling problems can be helped by the use of magnets to control the sheet both at the feeder, which is of a single sheet fed type, and through the machine. Aluminium is more difficult to feed since it does not respond to magnetic devices, retains imparted curl to a greater degree, and has a resistance to drag, making positive registration difficult. Control must therefore be light with minimum movement during registering and minimum impedence to the smooth and level travel of the sheet. Low tack conditions at the blanket surface are essential or sheets will pluck and travel round the blanket.

Register and colour work

Register and colour control are more simply carried out if plates are prepared with good surface selectivity and true tonal values. For example, colour control is not helped by having to starve the image of ink because non-image areas are insufficiently oleophobic, or because dot values are too strong. Makeready establishes the desired print target, the machine run has to maintain it. The most critical period for register and colour variation comes at the start of the run for the litho printer since correct ink/water balance has to be established at optimum running conditions, and work contrast tends to alter as print sharpens up with the gathering of momentum. Shorter machine runs are more vulnerable to this variability problem since makeready and start-up occupy a larger proportion of the run. Undercolour removal and good masking techniques help to ensure that an even film of ink is needed throughout the colour printings. This is easier for the printer to maintain.

Quality must be checked with target copies regularly during the course of the run, and the printer's judgement can be reinforced by off-the-machine checks with densitometers and register gauges. Register must be continually tested since feeder problems can easily arise through paper fluff or a mechanical defect without being immediately apparent. Printing surfaces, the plate and blanket, must be kept clean on the run or performance will vary. The prevention of troubles before they occur requires control of fountain solutions and ink duct settings. Variation in paper size will cause register trouble, and it is necessary to retain the same points of register through the job. Perfected work printed single colour requires a sidelay changeover when printed work-and-turn. Work-and-tumble perfecting requires paper which is accurately trimmed on both gripper edges, and for work-and-twist, paper should be trimmed on all four sides.

Litho colour sequences are normally decided by the need to avoid dot spread, register hazards, and overprinting problems. With single and two colour machines they are usually decided by viewing, to establish the weight of the design in the first printing. For nature scenes with predominant greens, blue and yellow would best be printed first; for browner scenes magenta and yellow could be the first pair. Presence of solids in a working sometimes cause a preference to print that colour later in sequence to avoid hickeys occurring through paper dust accumulating on the run. The first printing is the most vulnerable to dimensional change in the paper stock and a critical close register pass should be avoided for this printing unless the paper has been pre-stretched by a machine pass under pressure. Pale colours are more difficult to assess against the white background of a paper sheet and should not be chosen as the first printing on a single colour machine unless they need to be opaque inks.

Similar basic principles apply to printing sequences on multicolour presses. The critical problem of dot-spread will influence the sequence together with the need to trap succeeding colours wet-on-wet. Dot-spread occurs when rapid filtration of the liquid in the ink film becomes impossible. Previous full-tone printing, or the migration of polarised material from half-tone work in the first printings reduces the capillary attraction of the stock surface so that the superimposed ink dot migrates into the previous ink film instead. Black printed last, because it is darker, would more readily show a fuzziness affecting tonal values of screenwork. For this reason darker colours may be printed early in sequence on multicolour presses with transparent overprinting inks used to allow the first colours to come through as the ink film consolidates. Tack values must be adjusted to decrease sequentially so that following colours lift onto the preceding ink and are less prone to transfer onto the respective printing blanket. Contamination from preceding colours can adulterate the ink hue on the roller train, and is then likely to sensitise the other printing plates with secondary images which ultimately accept ink to print ghost colours.

Under ideal conditions darker colours should be printed later to hold their register more easily. Undercolour removal allows the black printer to play a greater part in producing the grey renderings and increases the necessity for it to be laid down later in sequence. High bodied, low tack inks with thixotropic characteristics are designed to reduce contamination and allow more flexibility in choice of ink sequence. Less structural strain is imposed on the paper surface with the lower pick value of such inks thus reducing the effect of this problem on register variation.

Colour consistency is not so easy to retain in lithographic printing owing to the need to maintain a balance between ink and water. The use of five and six colour presses will enable critical solid colours to be separated from colour halftone printing for some types of work. Perfecting units can be added to these machines; the perfector being the last unit in the system, to reduce register inaccuracies, and lessen the chance of smearing with print on both sides of the sheet.

Running letterpress offset

Photographically produced relief plates are increasingly being used to provide consistency of solid and line work on offset machines. There is no difference in principle to their use on letterpress machines. Peripheral speeds of forme rollers must be correct since there is less traction from a relief plate than from a lithographic surface.

Best printing results are obtained with light, accurate settings of rollers and pressures. This also reduces damage to compressible surfaces and this is significant where lithographic and relief plates are used alternately. When halftone is included in letterpress offset work, interference pressures are critical since any increase in impression will enlarge dot size in the highlights by a greater percentage than in the fuller tones and solids. Tonal values must therefore be maintained by standardising pressure and by accurate plate caliper over the whole surface, since any local packing for makeready is out of the question. Bearer contact presses provide the best traction control.

Post printing requirements

Most of the post printing requirements applicable to letterpress apply equally to lithographically printed work.

Some offset litho work has to be overvarnished after printing on paper stock. This is occasionally done as a machine pass, but more usually on spirit varnishing machines after printing. Inks used must be suitable for overvarnishing, and most lithographic inks are satisfactory.

Work printed on tinplate is nearly always overvarnished on special machines coating to fine filmweight tolerances. This may be achieved as an inline operation wet-on-wet in which case care has to be taken that the varnish does not dissolve the ink film. Bleeding into the varnish is a problem which has to be avoided where high stoving coatings are used to withstand food processing and canning. In this instance the printed film must resist the sterilising and retorting procedures in addition to the metal fabrication.

An aspect of fabriction, whether on metal or other materials, is that distortion of the printed design may take place. A drawn metal container has to be formed after printing, and it is necessary for the degree of distortion to be assessed to allow for this post-printing fabrication. The method requires

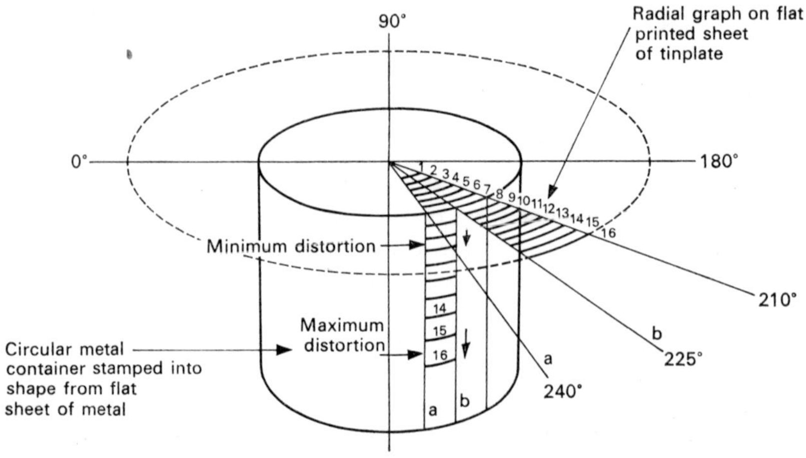

5.6 *Assessment of distortion on a circular container. A centrally stamped radial graph design is used to assess distortion on circular fabrications such as tin cans. Radial lines contract to become parallel down the drawn container as a–b. Elongation down the draw must equal the contraction of the circumference if metal substrate is to remain the same caliper. Distortion is worse further down the draw or on smaller diameters and the design has to be accommodated to this distortion.*

proving the required metal which is decorated with a graph design, and stamping this into shape so that the artist can draw his target decoration on the container. The design is then plotted onto flat graph decorated copy and the stamping procedure programmed to ensure that the degree of distortion is consistent throughout the run.

Transfers or decalcomanias are traditionally produced lithographically and require that the printing sequence shall be in reverse order to make the design right reading when the transfer is used. Transparent plastic materials are increasingly used nowadays – for example in the production of birthday cards – and require the same reverse order technique of printing. Copy can be viewed against a white or suitable background during the run and compared with colour proofs in the normal way. Set-off prevention may be by interleaving, and this is usually effected by supplying the sheets already interleaved and adhered to the interleaving sheets at the feed control edge.

Lithographic printing faults

The paper surface. Uneven absorbency of paper can cause mottle and spotty drying as well as contribute to chalking or isolation of the ink pigment on the paper surface. Ink dries patchily because paper surface conditions are not regular. Any interruption to the chemical drying of the ink film can result in spotty drying due to surface acidity, lack of oxygen, or local increases in humidity due to moisture evaporation. On coated or laminated stocks, press moisture can release latent acidity from the substrate, adhesives or backing. Press moisture fills the capillaries of locally poor receptive surface areas thus increasing the likelihood of uneven absorption of the ink film. Pre-testing of paper stocks with a suitable absorbency test avoids these pitfalls. K & N ink is particularly easy to use for this purpose by making a smear of the test ink over a range of sample sheets so that a comparison of absorbed colour will indicate any density variations.

Paper surface strength needs to be good for lithographic inks or rupturing will occur during printing. Paper coatings can pick up on the blanket surface, encouraged by wetting agents in the fountain solution, to cause piling which impedes image transfer. Additionally, fluffing is likely to occur as with letterpress and moisture helps to agglomerate the fibres to cause hickeys or plain spots in the printed image. Apart from correct choice and handling of paper stocks, minimum press moisture is the best precaution. Sometimes feedback from the paper surface or coating can sensitise the plate surface if a water soluble surface-active material is present. The scumming tendency may possibly be cleared by increasing the ink tack.

The printing surface. Washy marks in the leading edge of the printed image are a result of excessive press moisture. A reduction in colour strength which shows a mottled appearance under magnification is also a probable indication of too much water. Printing weight can be maintained with excess water by running excess ink, but print quality suffers. Sensible adjustment of supply will put matters right, bearing in mind the water-carrying capacity of the damping system in use. A seven-roll damper system with molleton covers takes much longer to compensate than a five roll unit with paper dampers. Overcompensation should be avoided since the opposite condition of scumming or catching-up will arise. Fine spray apparatus is useful for the local application of moisture to the dampers and is good, safe working practice.

Scumming can result from too little press moisture and an early opportunity should be taken to attend to the plate surface when sensitisation has occurred. Allowing the plate to dry during machine stoppages leaves a film of oil normally existing at the blanket interface to dry also, and removal of this film coupled with suitable densensitising procedure is always advisable. Scumming can also be due to faulty platemaking and can only be removed with a lot of local effort if photo-scum exists in the form of an image. Should the scum exist in the plate surface through excessive counter-etching or other cause, increase of fountain acidity and raising the ink tack will often clear the trouble.

Image blindness can also arise through platemaking faults, in which case a steady deterioration will occur. Washing out too clean when removing ink, running for long periods with dampers on and rollers off, or cylinder scuff due to excess pressure or surface speed differentials may cause or hasten image blindness. Ink should not be allowed to dry hard on the plate since removal then becomes difficult and efforts have to be made to restore the condition of the image. Photo-images should not be abraded when being reconditioned, as was the usual practice with transferred 'fatty acid' images. Instead a suitable agent should be used to soften the lacquer layer and create the right surface conditions so that the ink base will adhere.

If the right tonal values cannot be maintained, the fault may lie in origination failing to match screen values to press conditions. At low ink coverage tonal values are lost between fulltones and solids, whilst at heavy ink filmweight the highlight range is similarly affected. A good test of machine printing conditions is to use a 75% tone image with appropriate quality control strips and register marks on a grainless plate. Gear marking and pressure variations show up during such a print test and can be pinpointed.

Similarly ink filmweight must allow clear open printing at the same tonal value and platemaking must meet this target. Colour guides used to assist origination should be prepared under press conditions with the standard range of inks and stock if maintenance of correct tonal balance is required.

Similar factors cause trapping difficulties where inks being run are not balanced in strength for multicolour overprinting. Running a weak ink along with a strong one makes trapping difficult, especially if the weaker colour is first down. As with letterpress machine work, wet-on-wet trapping is dependent upon the physical aspects of adsorption and absorption, and control of materials to achieve this has been dealt with elsewhere. The influence of moisture on the surface tack of the press ink film is also a controllable factor which has to be taken into account when running. Increase in press moisture reduces surface tack of that ink, and consequently its tack balance with other inks used in sequence.

Faults occurring on the run

Apart from misregister through feeder faults or the expansion of paper with humidity variation, the litho printer may get local variation over his sheet. This is due to the alternate pull and freeing of tacky ink at the blanket interface where water lubricated, non-image areas come into the impression nip. This can be remedied by reducing ink tack as much as possible. The condition of the blanket surface is also important both for strip-off and impression quality. Ink, solvent and blanket washes should not be allowed to remain on the blanket or a combined oxidation with the rubber will result in a glaze over the surface and reduced transfer properties. The action of daylight can cause a variable characteristic of glaze and print as a ghost on subsequent work. A periodic cleaning with a suitable ketonic blanket deglazer will remove glaze. A wash with a damp cloth is also useful since some apparent glaze may result from the presence of gum and paper dust.

Contamination of the damping unit from press ink is unavoidable, and in some cases will affect printing performance. Tinting is the result of using an unsatisfactory pigment in litho ink which allows water soluble dye to leach into the fountain solution, with consequent tinting of non-image areas. Leaching-out of a non-water-soluble pigment and extender reduces the efficiency of dampers and may cause them to deposit ink on the plate. If the ink has high internal cohesion it can be removed from the damping system fairly easily. Low viscosity inks contaminate dampers more readily and are more likely to cause plate scum and damping troubles. Metal rollers in the damper train should be throughly desensitised and gum concentrate added to the fountain solution to retain this characteristic during the run.

Feedback of fountain solution into the ink roller train and ink duct causes poor flow characteristics to develop. Paper dust and size worsen this problem. Agitators are useful to shear the fountain ink and prevent it laying back from the fountain roller, and care has to be taken when refilling the ink fountain during the run. Ink density variation must be compensated for immediately to maintain print density.

Faults after printing

Ink drying and setting have been thoroughly covered in previous chapters but it is important on lithographic machines that press requirements should not cause secondary problems. Dry spray powders can be picked up on the blankets of following printing units resulting in plate scrub and subsequent wear. Use of acids to keep the printing plate clean will inhibit drying as will the use of excessive press moisture which emulsifies the ink.

GRAVURE

A large part of makeready in photogravure is carried out prior to a plate or cylinder being etched. It is important for the printing surface to be polished to a mirror finish in order that the doctor blade can remove all surplus ink from the surface. Correct size and accurate shape are important for quality and good register. Type matter for reproduction by gravure may be obtained from a number of systems. Reproduction pulls taken from type formes should have an even and consistent weight of impression and ink film coverage in order to reproduce well from the printing cylinder. Photographed and photo-composed type matter should be quality controlled at all stages of production in order to ensure facsimile reproduction of the original face. Type matter should be visually checked by an expert, and where possible tone patches or step wedges used to enable measurements to be made with a densitometer. Any deficiencies or defects will not disappear but are more likely to be amplified.

For monochrome and colour reproduction emphasis must be placed on good camera work and the utmost detail to colour separation and retouching. In order to obtain some assessment of the separations prior to committing the work to the cylinder various methods are available. Optical systems using a viewing screen and colour filters will project a coloured picture onto the screen by superimposing the separation positives in sequence next to one another. Other methods for checking are photomechanical or electrostatic proving systems which provide a complete coloured picture for assessment by a responsible person. Once the proofs have been checked and approved by the customer, planning takes place. Accurate and careful positioning of the text and pictures is of paramount importance, for once the cylinder has been etched it becomes impossible to move individual pages in order to establish correct register.

In many cases cylinders and plates are proved on the press which will ultimately print the job. This is an ideal situation but often a rather expensive one. Proving presses are available either as a single unit sheet-fed device or a multi-unit web-fed press, usually consisting of four units. Press-room inks are suitably modified in order to simulate actual press conditions and proofs obtained. Any modification to ink will be made and all the details recorded, a further proof taken and submitted to the revision department. When cylinders have been revised they may be returned for further proving or made available to the pressroom with any relevant recorded data.

The actual makeready sequence for gravure sheet-fed machines is similar to letterpress and lithography with a few exceptions which are particular to gravure. Lubrication and cleanliness are very important as liquid inks flow and splash quite easily and will cause a bearing to seize if not given attention.

In the preparation of a printing plate it is necessary to ensure that the back of the plate is clean and smooth as any particles of dirt will cause a surface pimple and a subsequent printing defect known as a 'butterfly'. Some press designers allow for a thin card to be inserted behind and to cover the whole area of the plate; in the event of dirt getting behind the plate the card will absorb it. However should a creased sheet pass under impression while printing there is every likelihood that the plate will be battered.

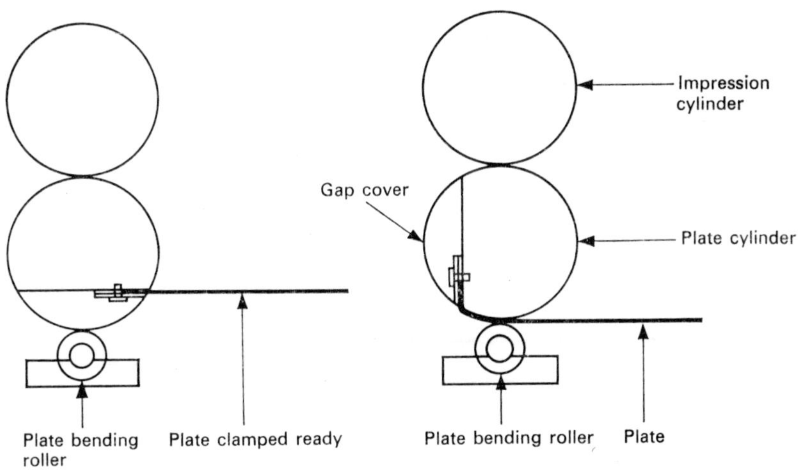

5.7 *Plate bending on a gravure machine. The plate is locked into position on the machine and then turned into the impression nip where the bend is formed.*

Plate bending can be carried out on the machine or off the machine. On the machine the plate must be centred on the cylinder and securely clamped, a rubber roller of about 90 shore placed in position beneath the plate cylinder and the machine moved round in order to bend the plate to the cylinder. The tail is then secured in the back clamp and the plate tightened down.

The part of the plate cylinder which is not curved houses the clamping mechanisms. These are covered with a segment or gap cover which is bolted into position after the plate has been run on impression to ensure that it is properly bedded down. The bends on plates should not be sharp; a torn edge may give rise to nicks in the edge of the doctor blade and it is not unusual to burnish along the bends in order to avoid possible troubles. A perfect ink seal is obtained by the use of grease to prevent ink seeping under the gap cover.

Centre lines scribed or etched into a plate will assist in the accurate positioning of the plate on a cylinder bearing related scribe lines. Large plate machines are fitted with identical clamps at either end of the plate, making it easy to swing the plate should it be necessary to obtain register.

Plate bending machines are supplied with some presses and consist of a platen, clamping and bending device and two front marks to which the sheet

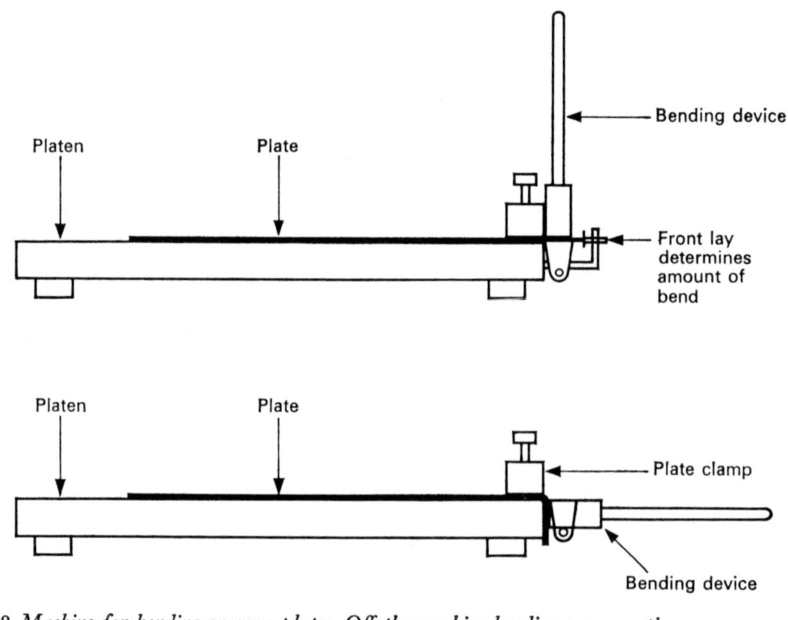

5.8 *Machine for bending gravure plates. Off-the-machine bending can save time and most manufacturers will recommend a machine which will form the bend with accuracy.*

of copper is located then clamped and bent. The two front marks are adjustable and determine the amount of copper to be inserted in the clamp. The plate is turned 180° and the bend aligned with the back edge of the platen; the tail is secured by the clamp and then bent. This is an important operation because if the bend is fractionally out the register error will be larger on the tail end of the plate once it is on the press. Fine register on sheet-fed machines is obtained by lay adjustment; it is essential, therefore, for the image to be laid down and etched on the plate in a position determined by the allowance for clamping, the pitch line of the impression cylinder, and a centre line for lateral positioning. If this has been done accurately only micrometric lay adjustments will be necessary. Provision is made in plate cylinder design to advance or retard the cylinder in relation to the drive gear by releasing the cylinder retaining bolts; in this way initial circumferential register is obtained.

The impression cylinders on sheet-fed machines are engineered to allow for a packing. Sheets are inserted under a rubber blanket to make the circumference of the impression cylinder compatible with that of the plate cylinder. Since bearer contact is not a feature of sheet-fed gravure presses the sheets are staggered at the gripper edge and at the trailing edge to allow the impression cylinder to run on and off the printing surface without bumping. Bumping can produce a printing defect known as a 'kick mark' caused by the doctor blade vibrating on the image and causing it to wear.

With sheet-fed gravure printing the temporary sheets of packing permit local pressure adjustments of a general nature, or local treatment by way of flush sheets, mounted on a makeready sheet and registered in the temporary packing. This increase of pressure in the image areas results in better ink transfer from the cells and the non-image areas run cleaner for a longer period because of the absence of pressure. The paper which comprises the packing must not be too soft or it will compress with consequential loss of pressure. On the other hand some compressibility is desirable with a hard rubber blanket. Generally speaking, the harder the packing the sharper the image, and the more pronounced will be the inequalities in the cylinder or packing. A compromise is therefore needed, taking into account the machine, the job paper and the shore hardness of the rubber blanket.

Because bearer to bearer contact of printing and impression cylinders is not normally possible in sheet fed machines, provision is made for overall pressure settings by means of set screws acting on cylinder journal bearings. Calibrated pressure settings are an advantage, but with many machines it is necessary to rely on test strips of paper being placed in the nip at either end of the impression cylinder. The pressure screws are adjusted until each is held with equal firmness.

Pressure settings should be kept to a minimum and register established by movement of the cylinder, leaving the lays for fine adjustments. Other makeready procedures are checking and establishing the correct viscosity for the ink; filtering it ready for use in the inking system; setting the inking system to ensure a sufficient supply of ink to the cylinder; and adjusting the doctor blade to provide a clean wipe at a minimum operating pressure. On printing units where a blower pipe is used to control background tone or the tonal result of the etch, this should be clamped in position and given a provisional setting. The application of heat for drying purposes should be kept to a minimum to avoid distorting the substrate.

Makeready on web-fed machines

On web-fed machines there are no facilities for making local pressure adjustments, and in multi-colour work, if perfect register has not been achieved at the stage of printing surface preparation, there is nothing the machine printer can do other than offer a compromise. The main operations are putting the cylinders into the units and thoroughly cleaning them; leading the web in its correct path through the machine; setting the pressure units correctly and establishing initial register. All register rollers should be centralised and the press prepared as for running, and ideally the register roller assembly should form an 'n' with the moveable roller at the top, where it exerts the maximum influence for the minimum movement.

There are various methods employed for the initial registration of cylinders in a multi-unit press. The first is to print all cylinders on the web, establish one as the key printer and advance or retard the others to fit the key; another method of registering is by using an ink spot, which is put on a clearly identifiable registration point on the first cylinder and as the web passes through successive units the press is stopped and the cylinders adjusted to the register spot. The cylinders can then be inked, doctor blades set, the press started and final register obtained by using the register rollers when the press is at production speed. The same principles can be applied for initial register on machines where differential gears are used for registration. Lateral register is obtained by the sideways movement of cylinders and to obtain quick registration it is important that in planning and etching cylinders all images are centred laterally.

Whenever practicable all cylinders should bear a common registration mark which can be aligned with a suitable mark on the printing unit to establish initial register, and with variable size machines a record kept of the position of each register roller together with the cylinder size.

Once printing register has been established the process can be carried on to the folder, which can be declutched in order to register it with the print. Final folder register will be established with the register rollers. Control of multi-colour printing is aided by register, slitting and folding marks, and step wedges will indicate the tonal range required.

Screen angles and colour sequences in gravure

For monochrome printing the normal screen angle is 45°. It is optically suitable and undergoes the minimum stress from the doctor blade. A 90° angle should be avoided because it is subject to physical stresses from the doctor blade which may cause the image to break down.

In conventional gravure multi-colour printing it is usual for the screen angles to be arranged at 45° for all colours. With some substrates however, it is necessary to consider giving each colour a different angle in order to avoid 'screen clash' and 'colour shift'. These are caused by the ink from the cells forming regular patterns and are most noticeable in the middletones and highlight areas. The slightest variation in the superimposition of one colour on another will bring about a colour change and screen patterning.

These two problems also occur with inverted halftone and electronically engraved images and may be resolved by the juxtapositioning of the cells with suitable screen angles. If however they do occur when printing, it may be possible to cause the ink to flow on the substrate sufficiently to prevent any regular pattern from printing by carrying maximum solvent in the inks. Some firms use a mixture of inverted halftone and conventional gravure images in the production of their colour work with a great deal of success.

Colour sequences on multi-unit presses are normally yellow, magenta, cyan and black, or light to dark colours, and these are rarely altered. However when printing on transparent materials for reverse reading the opposite order is used.

Gravure printing faults

Streaking. A number of streaks appear on the copy from around the cylinder. Their positions will change with the oscillation of the doctor blade. The following are the causes:

1 Paper dust or coating trapped under the edge of the doctor blade. In order to minimise or prevent this from happening suitable brushes or vacuum boxes can scavenge the web just after it leaves the reel. However, if the ink is too thick it will pull the surface of the paper in the nip, in which case care must be exercised to maintain correct viscosity.
2 Ink drying under the doctor blade, sometimes referred to as 'beading'.

Check for correct ink viscosity. This may also be ink agglomerates, in which case it will be necessary to consult the ink chemist.
3 A fault of the doctor blade due to insufficient stoning and polishing after having prepared the initial edge, which leaves it rather ragged.

Tone band or trellis. This printing fault is a band of tone which, when fully developed, is the width of the doctor blade traverse and goes around the cylinder. If the blade oscillates slowly a toneband is produced; if it has a rapid oscillation then a trellis is formed. This fault must be dealt with quickly as the matter lodged under the doctor blade is hard enough to score the cylinder. The best prevention, therefore, is cleanliness. Polishing the band with fine emery polishing paper and stoning the edge of the blade where the band is occurring may resolve the problem. Failing this a new doctor blade should be tried.

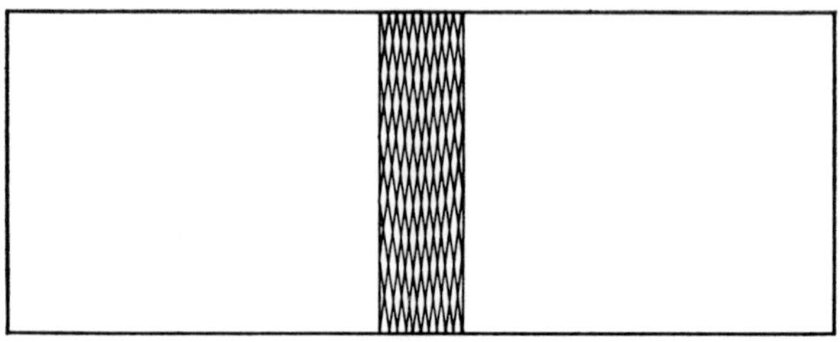

5.9 *Gravure tone band. This pattern is cut into the cylinder corresponding to the width of the doctor traverse. Quick action is required to save the cylinder from permanent damage.*

Persistent lines are parallel lines appearing along the web or across the sheet in the machine direction and moving across the copy with the reciprocation of the doctor blade. They are the result of nicks in the blade caused by hard matter embedded in the cylinder. Here again cleanliness is important as a preventative. A pickfinder can be held against the cylinder in order to remove the source of the trouble.

Comet or arrow. This fault is caused by a spur of copper, chrome or foreign matter projecting from the plate or cylinder which nicks the doctor blade, causing it to cut the printing surface for a short distance until the nick wears out. At the next revolution the blade is nicked again and, as it reciprocates, produces a series of fine lines fanning out in the shape of a comet. The spur can be removed with a graver or the surface burnished.

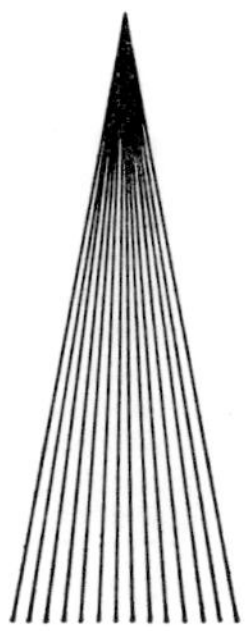

5.10 *A 'comet', caused by the doctor blade being nicked and the nick cutting the plate or cylinder, wearing out and being nicked again on the next revolution. Doctor traverse accounts for the pattern.*

Gear marks are liable to occur on older machines. They appear as light and dark stripes across the print, related to the pitch of the gears. This generally occurs when the gears are too deeply enmeshed, although it has been known to appear when gears are insufficiently meshed.

Kick marks have been referred to previously as a variation in print strength across the cylinder due to the doctor blade pressure changing. This is usually caused by a vibration such as the cylinders bumping on and off impression and the fault will be accentuated if the cylinder journals are worn.

Mottle is caused by an excess of solvent in the ink. This may be caused by poor monitoring and control of viscosity, or by sudden dilution of the ink with neat solvent causing a temporary unequal distribution in the trough.

A 'dry' print is the result of too little solvent and again would be the result of poor viscosity control. This can be recognised by broken or missing high-light tones and the loss of smoothness in the solid areas.

A butterfly as mentioned earlier in the chapter is caused by a particle of dirt beneath the plate, pressing into the copper to cause a surface pimple. When the plate is inked and wiped by the doctor, two spots of ink will be left at the sides of the pimple, like a pair of butterfly wings. If the fault occurs in the image area the plate must be returned to the process department for repair, but in a non-printing area it may be removed by a sharp chisel then polished, or the packing on the impression cylinder may be cut away.

6. Web printing

WEB CONTROL

A reel is unwound by suspending it in the machine in such a manner that it may revolve, and then taking the free end of the paper and pulling it towards the printing units. The revolving action may be accomplished by passing a spindle through the reel core and supporting the spindle ends in bearings. Such an arrangement is called a 'reelstand' and can be a separate or integral part of the machine. The pull on the reel can be performed by the printing cylinders or a specially designed pulling unit geared directly or indirectly to the machine.

To this basic concept must be added a control system in order to 1, govern the peripheral speed of the reel in relation to the machine; and 2, maintain a constant tension in the web so that print register is maintained and faults such as creasing, slur and web-breaks do not occur. These two requirements are usually interdependent.

If the reel is unwound by the pull from the machine, then the tension in the paper will only equal the force required to overcome the inertia of the reel and the friction at the spindle bearings. The revolving reel will gain in momentum as the machine accelerates, and upon deceleration of the machine the inertia of the reel will cause it to continue to turn freely so that the relationship between machine speed, web speed and reel circumferential speed is lost. To prevent this a braking force is applied to the reel to prevent overrunning and maintain a constant tension in the paper under all conditions.

Relationship between momentum and braking force

A reel of newsprint with a diameter of 950 mm, weighing 750 kg and revolving at a rate to match a web speed of 10 metres per second will have a momentum of 7 500 kg/m/s. The force required to stop this reel in an emergency, such as a web break, would be comparatively large, *eg* a counterforce to stop the reel from turning more than one metre in one second would need to approach 7 500 kN. For web control while running, however, it will only be necessary to use sufficient force to induce tension in the paper, while keeping well within the elastic limit and breaking strength of the particular stock.

As a rough guide, a counterforce of less than 0·1 per cent of the force required to stop the reel would have little effect on most paper stocks, while a braking force of more than 1 per cent will risk permanent stretching and possible web breaks, *eg* in the above instance a retarding force of 7·5 kN would be negligible, while 75 kN would be excessive.

The need for continual brake adjustment

It is not possible to give set figures for web tension control because one of the first principles of reel braking is that the retarding force must be constantly adjusted due to the steady reduction in size of the reel as it unwinds. If brake is to be proportional to momentum, then at any constant web speed a continual reduction of braking force is necessary as the reel size reduces. Similarly, speed variations will affect momentum and require brake adjustment. A further complication is introduced by the fact that although the surface speed of the paper is always related to machine speed, the number of revolutions per minute will increase as the reel gets smaller.

Methods of braking

Straps. The simplest way to control web tension via reel braking is to suspend a weight upon the reel by means of a strap. The strap may be of leather, webbing, plastic or metal, and the retarding force will be direct friction between the strap and the periphery of the reel. The efficiency of braking in this instance will be influenced by the mass of the weight suspended, the

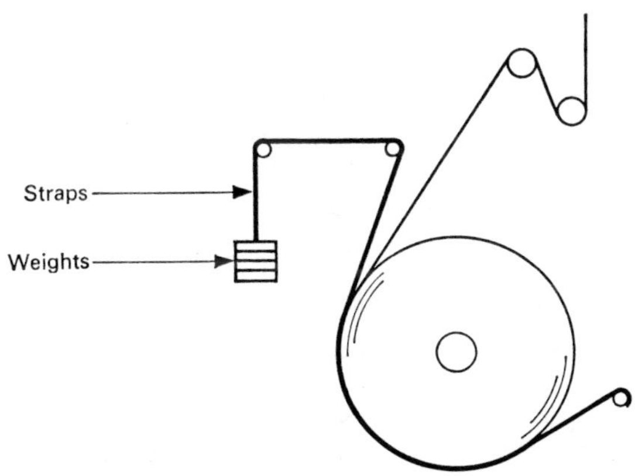

6.1 *Strap brake. Tension is applied directly to the surface of the reel by means of straps and is regulated by weights or some other force applied to one end of the straps.*

area of contact between strap and reel, and the co-efficient of friction between the two surfaces. A reduced weight will be necessary as the reel becomes smaller, but this is qualified to some extent by the reduced arc of contact, so that generally speaking the efficiency of the brake is reduced as the reel runs down.

Function of the tension roller. As the aim of reel braking is to produce a state of tension in the paper, adjustments to the brake must be made by reference to a sensor in the web path. The sensor in its simplest form is a roller, so arranged that the pull between machine and reel will move the roller from its resting position in proportion to the tension in the paper. Brake adjustment can then be made by visual reference to the position of the tension roller, or in a complete control system the tension roller itself will be linked to the brake mechanism so that automatic adjustment is made. This will ensure that not only is the web tension adjusted to reel diameter, but speed changes and emergency stops can also be rapidly compensated.

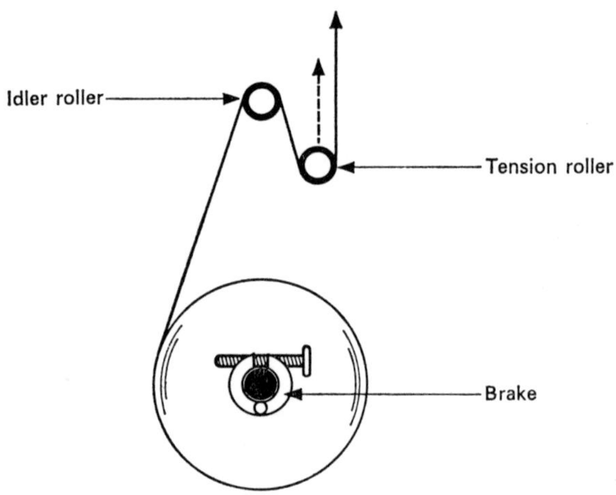

6.2 *Core brake and tension roller. Tension is induced in the web by a pair of brake shoes acting on the end of the reel spindle. The tension roller is free to move in an upward direction from its resting place. The position of the tension roller will depend on the resistance offered to the pull of the machine by the brake.*

Choice of force. Instead of weights an electric motor may be used to pull at the straps. Other alternatives are to control strap pressure by pneumatic or hydraulic systems. These alternatives provide an easier linkage between the tension roller and braking force for automatic control than the mechanical linkage necessary when weights are used.

Running bands. Strap systems are the simplest of brakes but they are often responsible for marking the paper surface, particularly with coated stocks, and for static electricity charges in the web. An alternative is to use a running band to control the reel.

This system is basically an endless belt connected directly or indirectly to the machine drive and running on the periphery of the reel at a slightly slower speed to induce web tension. Again there are two principal factors affecting the efficiency of this brake: the area of belt in contact with the reel and the difference in surface speeds. Where the running bands are connected directly to the machine drive they will be geared to run slightly slower than the printing cylinders and tension adjustments will be achieved by varying the arc of contact between band and reel. The alternative is to drive the bands through a variable-speed gearbox, so that tension variations can be made by slight changes in belt speed.

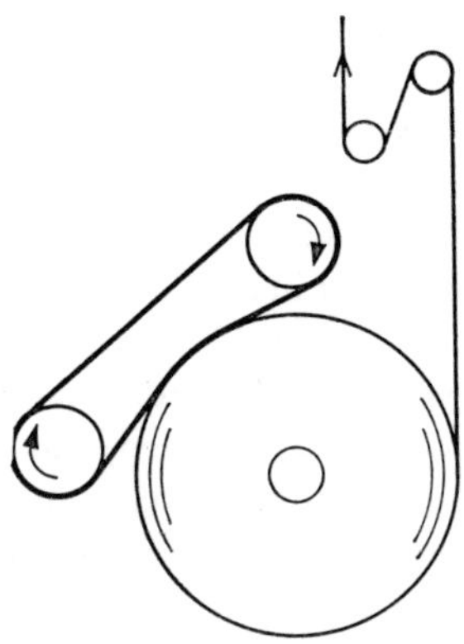

6.3 *Running band brake. The endless belt runs in contact with the reel, but is less likely to mark the paper than stationary bands. Tension can be regulated by varying the belt speed or the pressure of the belt on the reel.*

Core brakes. Where it is required that no straps, either stationary or running, shall touch the paper surface, a third type of brake may be used which operates through the reel core instead of its periphery. This can be a simple strap or a shoe brake gripping a boss at one end of the spindle. The brake can be controlled by manual adjustment or by a built-in link to the tension roller. As with strap brakes, electrical, mechanical, hydraulic or pneumatic linkage can be used between tension roller and brake. Unlike peripheral brakes, core brakes are more efficient on smaller reels than large ones because they are less able to control the centrifugal forces associated with large reel diameters at high speeds.

Direct motor control from the reel centre is also possible, bearing in mind that a stepless, variable-speed motor would have to be used because the core must revolve progressively faster as the reel becomes smaller. A motor driving a reel spindle would need to be independent of the machine drive, and in order to maintain a constant tension in the paper it must incorporate a speed control device related to the position of the web tension roller.

Web control within the machine

Web control starts at the reelstand. Every part of the machine which touches the web also controls it to some extent. Of particular importance are the printing nips and the feed rollers, of which there may be several sets on a large machine. If the surface speed of any two of these moving parts should differ then so will the web tension between them. This can result in slack or overtight paper, with all the troubles associated with these faults. In some instances however, particularly where long web leads are involved, it may be considered good practice to increase slightly the surface speeds of the web control members, progressively through the machine, so that tension is maintained between the units.

It is possible to unwind the reel by a direct pull from the printing unit. A firm and positive grip of the paper in the unit is essential for this system to be successful as the pull from the brake is in direct opposition to the pull of the printing nip. This will work on the arch-type perfecting unit used for publication work where metal plates are used, and control is aided by a good wrap-round of the impression cylinders. Accurate packing of the cylinders is important here to ensure correct circumferential speeds.

In principle all cylinders in a press should be of the same circumference in order to feed the same print length from unit to unit; this places the emphasis of web control on the pre-printing or post-printing operations and in some cases pulling groups positioned along the press. An alternative method is to increase the circumference of cylinders progressively from the first colour to the last. The amount of increase in circumference is a matter for experiment: 0·5 mm on the circumference of each cylinder may provide good web control between the printing units while an excess of this amount may lead to web breaks. The moisture content and elasticity of the substrate will also affect the performance where this method is used.

A fundamental difference occurs in web offset printing with the mechanical stress imposed in the impression nip. Contact between surfaces is not continuous since a gutter exists where plate and blanket ends are clamped. A release of tangential and impression build up occurs at this gutter and causes a fluttering of the web when tension is slack. As reels are printed with the grain running in the machine direction, the paper structure is weaker across the machine and creasing becomes more likely than if the grain were possible the other way. Build up in advance of, or behind the contact line of one impression nip alters the tension relationship at the adjacent unit on a multi-colour press and consequently alters register between the two printed colours. As a result it is usual practice to maintain higher web tensions for offset machines than with direct printing systems.

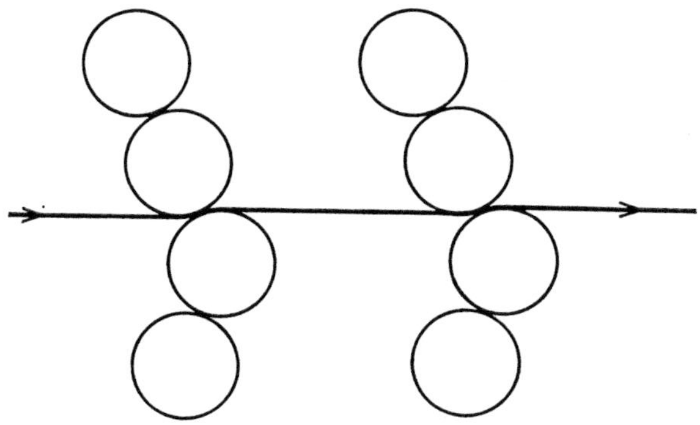

6.4 *Web flutter, caused by release of the web when gutters meet in the nip, is minimised by increasing the wrap of paper round the cylinders near the contact line.*

Most offset press manufacturers employ bearer contact between plate and blanket cylinders to eliminate bounce at the gutters and use surface settings between two contacting blankets. Uniform pressure between blankets on all units is needed to keep even tension on the web through the machine. Additionally, in order to retain good register between colours, multi-colour machines should have their units packed progressively as on sheet fed presses. This involves printing the first colour small by overpacking the plate and under packing the blanket to compensate. Sequential units require progressively less plate packing and more blanket packing to compensate for paper stretch. Accelerating peripheral speeds in succeeding pressure nips also helps to keep the web taut between units.

Paper-feed units are used for direct or offset machines where more accurate control of the web is required. In its simplest form a feed unit is a driven roller geared to the circumferential speed of the printing units. The web runs over this roller and a positive drive is obtained by a set of wheels pressing the paper on to the driven roller. For finer control a variable-speed drive may be used to govern the speed of the feed unit. In this instance the influence of the reel brake is only significant as far as the feed unit, which in turn controls the tension of the web travelling to the machine. Further feed units may be used to obtain complete control of web tension after printing and before the finishing operations such as folding or rewinding.

Registering on a web

Web register is the relationship between the web path and the printing cylinders. Providing that the cylinders or plates have been accurately planned and mounted in the machine there are three types of adjustment which may be necessary; length register in the web direction, lateral register across the machine and helical register which combines both movements. Either the cylinders or the web may be moved and some machines have provision for both types of adjustment.

Lateral register is obtained initially by centring the web and the cylinders in the machine. The printing units, or individual cylinders are usually provided with a means of sideways movement to allow for slight adjustments to their positions while the machine is running. Problems may occur where the web is subjected to distortion, due to the stresses imposed as it passes through the printing units or the effects of heat in a drying system. Where distortion is known to occur its effects may be reduced by passing the web through a pre-conditioning or pre-stressing unit before the first printing. If this is not sufficient to compensate for the distortion and a progressive variation occurs, it may be necessary to make proportional allowances for the register difference at the planning stage.

Longitudinal register. On many machines the printing cylinders can be revolved independently of the rest of the machine to obtain the correct relationship to each other and the substrate. Where the cylinders cannot be individually rotated, the machine is probably designed to produce a fixed-size product with the web path and cylinder positions in fairly constant positions. In either case the initial register, established during makeready, will need to be modified accurately during running due to the dynamic stresses incurred by the substrate and printing cylinders. One method of achieving length register is to employ differential gearing in the cylinder drive so that cylinders may be independently rotated relative to each other while the machine is running. Other factors affecting register are related to the rollers over which the web runs between the various units.

Idler rollers are the free-turning rollers which serve to guide the web through the machine. In addition to controlling the web path they exercise control over web tension by their relative locations and the degree of wrap-round. Their positions also determine the length of web between printing units and other operations; the choice, therefore, of web path will affect register between units.

Register rollers serve the same purpose as idler rollers in guiding the web through the machine, but they are arranged to be moveable within a certain arc so that the length of web between units can be altered. The movement of register rollers can be done while the machine is running so that even the smallest variations in length register, due to speed changes or tension fluctuations, may be corrected immediately they are detected.

Slewing rollers. If an idler or register roller becomes maladjusted so that it is out of parallel to the printing cylinders it will cause uneven tension across the web and move the web from its true path, as well as affecting the register. Some machines incorporate rollers which may be deliberately slewed either to move the web laterally across the machine or to achieve a slight helical adjustment to the register. As the slewing adjustment will also affect lateral web tension these rollers are best used in pairs. A pair of slewing rollers may be used either side of a printing unit for minor helical register adjustments, or as a composite pair to correct lateral web wandering.

Bowed rollers may be used to level out unevenness in a taut web which may

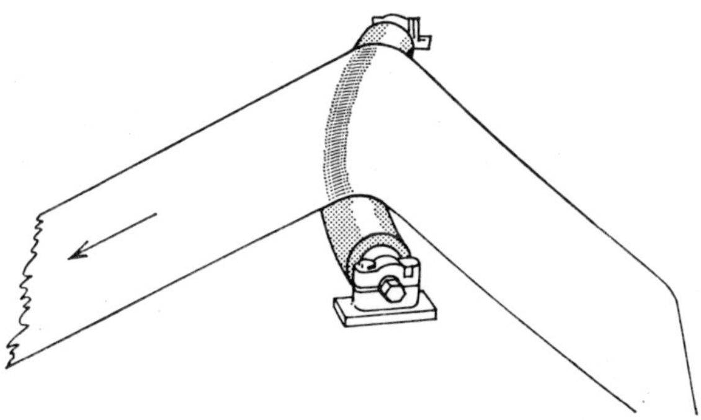

6.5 *Mount Hope curved roll (exaggerated). Materials which are prone to distortion in web form may be satisfactorily controlled by inducing a deliberate, predetermined distortion by means of a slightly bowed roller which will ensure an even, tensioned traverse through the machine without creasing, folding or buckling.*

lead to creasing or wrinkling. These are designed with an inner, stationary axle having a controlled bow and an outer sleeve which rotates on ball bearings.

Serrated rollers are used to reduce the problem of marking by wet ink passing over rollers. Contact between web and rollers is restricted to the peaks of a number of fine serrations in the roller surface and any of the previous types of rollers may be designed in this way.

Web turning

There are occasions when the direction of travel of a web needs to be altered for the convenience of folding, perfecting or associating webs. Three instances are given here.

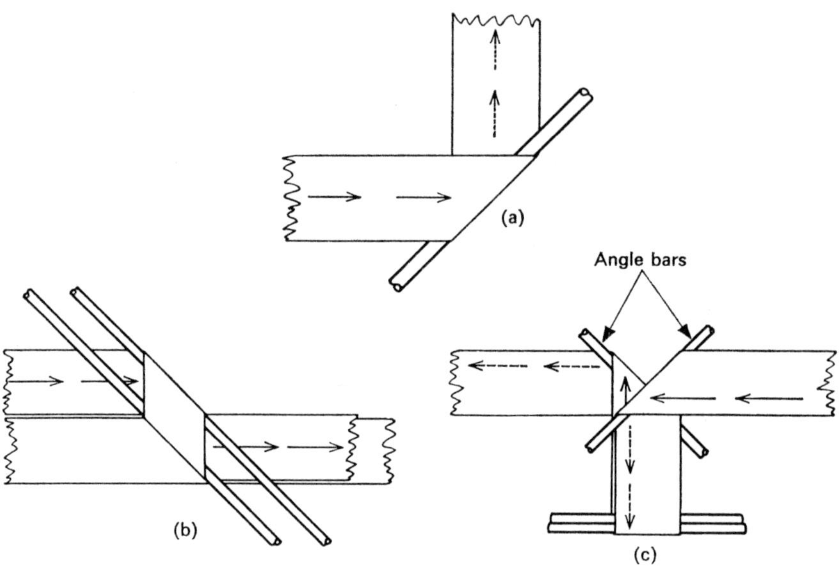

6.6 *Web turning.* (*a*) *An angle bar. The bar is placed in the web path at a 45° angle. Wrapping the web around the bar changes its direction by 90°.* (*b*) *A turner bar section. The web, coming from the left is slit into two half-webs. The arrowed half is wrapped around the bars and moves across the machine to a position immediately above the other half, which has passed under the bars. The two half-webs then proceed to the folder.* (*c*) *Web turnover section. The web first passes over a 45° angle bar, which turns it at right-angles. It is then turned over again and directed back to the machine by one or two free-turning rollers. A second angle bar turns the web over for a third time so that it regains its original direction.*

An angle bar is used to turn a web through a right-angle from its original direction of travel. This is a fixed bar, placed in the web path at 45° to its direction of travel. The web is wrapped round the bar, and in doing so changes its direction by 90°. To avoid marking as the paper rubs over the stationary bar, low-pressure air may be blown through sets of holes in the bar to form a cushion of air between it and the paper.

A turner-bar section comprises a pair of angle bars so arranged that half of a slit web may be wrapped around them. The paper thus makes two consecutive 90° turns, the result being that it is moved across the machine and continues to travel in its original direction in association with the other half of the original web.

A web-turnover section as its name implies is used to turn a web over to its reverse side for perfecting. This section comprises two angle bars at right-angles to each other plus a roller in line with the normal web travel. The paper is in fact turned over three times so that it is finally the opposite way up but travelling in the same direction as its original line of travel. A similar system may be used for double-passing the web, so that it may be taken back and run alongside a later section of the same web performing the previous pass. As this system is a perfecting method an oven pass is usually included in the turnover path and problems may arise through differing tension requirements between the two parallel webs running through the same printing units.

Reel handling

On the simpler machine at the expiry of a reel the core must be removed from the machine and de-spindled. A fresh reel is then brought to the machine, the spindle inserted, and lifted on to the reelstand. The leading edge of the new reel is then glued or taped to the tail of the web.

Various types of spindles are used, the most common comprising a shaft with two cones sliding along it. The cones are forced into each end of the reel core and locked in place. A more efficient system is to employ an expanding mandrel, which is not only quicker to insert into the reel, but kinder to the core. The mandrel may be expanded by internal sliding wedges or pneumatic pressure.

Spindleless systems may be used where the reel widths are fairly constant. Here a pair of cones are fitted into the reelstand sideframes instead of on to a spindle. The cones can be inserted or extracted very quickly by a simple mechanical adjustment. Spindleless reelstands are particularly useful with very big reels as they avoid the necessity and inconvenience of extra-long shafts.

The reel changing operation may take only a few minutes. This is acceptable for short-run work or comparatively slow production, but where maximum efficiency is required this downtime must be reduced to the minimum.

Double stands. One way of reducing the time lost through reel changing is to

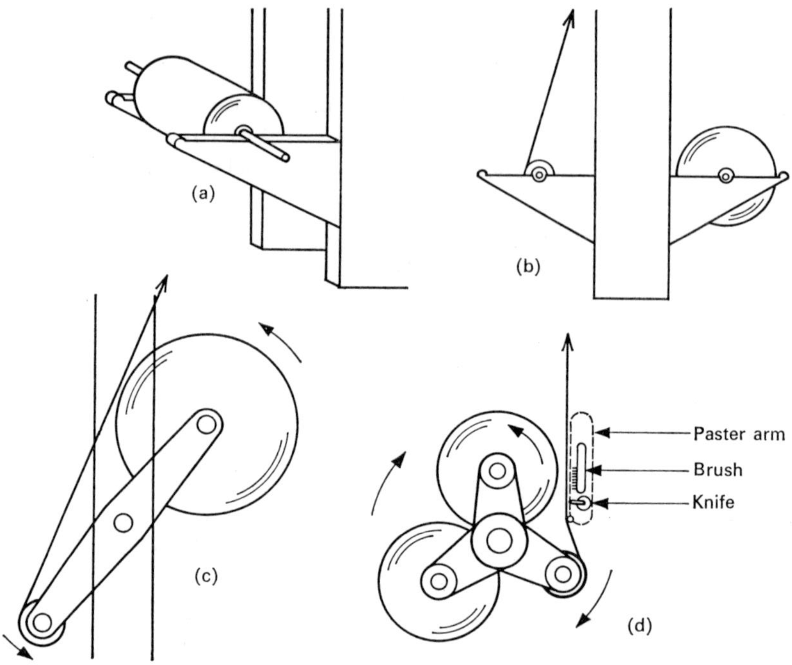

6.7 Reelstands. *(a) Single reelstand. The spindled reel must be lifted on to the reelstand manually or by a hoist. (b) Double reelstand. A second reel can be lifted into position and prepared while the first is still running. (c)* Revolving *reelstand. The arms are normally in a horizontal position when running, but the illustration shows the unit partly revolved and ready for splicing. Before the lower reel expires the web can be pressed on to the glued edge of the new reel while the machine is travelling slowly. (d) Automatic reel changing. The illustration shows a revolving reelstand with three arms, in the pasting position. The web from the expiring reel is running a few millimetres from a full reel. The brush will press the web on to the glued edge of the new reel and the knife will cut out the old web.*

provide a second position at the reelstand so that a fresh reel can be prepared and loaded while the machine is running. Upon stopping, the tail of the web can be attached to the new reel in a matter of seconds.

Revolving reelstands. A better system is to arrange the two (or three) arms of the reelstand to revolve about a midway axis, so that the prepared reel can take the place of the expired core and vice-versa. Other advantages of the revolving reelstand are that reel lifting is done by the reelstand arms, thus avoiding manual loading or reel hoists, and it is possible to make a 'splice' while running slowly.

This last statement needs explaining. It will be obvious that if a new reel is prepared while a web is running in the machine and the reelstand revolved, the reel must eventually touch the web. The operators take advantage of this fact and prepare the leading edge of the reel with a special adhesive which will not dry quickly or penetrate right through the paper. This glued edge will adhere to the web when the reelstand is rotated, and to avoid snatching the reel is carefully revolved to match the web speed before contact actually takes place. Needless to say this procedure can only take place at very slow web speeds when human judgement is in control, but the same basic principles are employed in the design of a fully-automatic reel changers for use at top speeds.

Automatic reel-changing devices

Most automatic reel-changers are based on the principle of the revolving reelstand. The reel is prepared with a glued edge for presentation to the expiring web and the following basic steps are carried out.

1 Revolving of the reelstand to ensure a precise relationship between the prepared reel and the web.

2 Acceleration of the reel to match the web speed prior to joining.

3 Splicing – the actual mating of the web and reel to make the join.

4 Breaking-out of the expiring web once the join has been securely established. Tension control must be practised during the whole of the splicing cycle.

These items are now dealt with in more detail.

Revolving the reelstand. This is done by motor control and can be started by the operator or by a measuring device on the running reel to start the splicing cycle at a pre-determined diameter. The distance between the reel and the web prior to the join usually needs to be about 10 mm–15 mm, and for accuracy must be established by machine control. A simple way is to allow the reelstand to turn until the reel interrupts a light beam just in front of the web. The light beam in turn controls a relay via a photo-electric cell. An alternative method is to limit the travel of the revolving reelstand by reference to a measuring device resting on the reel which controls the amount of movement according to reel diameter.

Acceleration of the reel. Driving bands may drop onto the reel once its position has been established. These can be either a special driving belt designed for this purpose, or the running bands which also control web tension.

Other systems drive through the centre of the reel, using a variable-speed motor or a mechanical drive through a positive infinitely-variable (PIV) gear system. With the Cline system the reel revolves by contact with the moving web under pressure from running bands. In this instance the adhesive on the leading edge of the reel is inactive until sprayed with an activating solvent.

Splicing. Most systems employ a brush which comes into operation just after the prepared edge has passed the point of contact. This presses the web on to the reel for one complete revolution to ensure full contact. The Cline system splices by spraying the adhesive with a solvent while the web and reel are in contact so that an immediate join is made.

Breaking-out of the expiring web is invariably done by a knife which extends the full width of the reelstand. This is timed to come into operation approximately one revolution after the splice is made, and will also work in conjunction with the brake on the expired reel to stop the core from unwinding the remainder of its paper unnecessarily.

Tension control throughout the splicing cycle applies to the expiring reel and the new one, particularly for the short period while both reels will be running together. Where peripheral brakes are used for the main tension control an auxiliary centre brake may be added to control the core during the changeover. This is a satisfactory system in the circumstances, bearing in mind that the centre brakes are most efficient on smaller reels, while peripheral brakes are better for larger reels. Where tension is controlled by motors driving through the centre of the reel a separate motor is required for each arm of the reelstand. This is certainly a more expensive method, but it does ensure continuous control by a single braking device on each reel and reduces the possibility of tension variations during changeover.

Any automatic reel changing mechanism must necessarily be a reliable system. If a splice should fail at full speed the web will probably pass right through the machine before the machine can be stopped. The time lost in threading the web and incidental work such as cleaning the blankets will take longer than several manual changes.

Folding a web

There are two basic types of folds for webs; 1, folds in the machine direction (length folds); and 2, folds at right-angles to the machine direction (cross folds).

Length folds are made by wrapping the web around a shaped forming plate and at the same time pulling it across the plate. This 'former' is usually a triangular or conical shape where a simple length fold is required down the centre of a web. For multiple and non-central length folds, as used on paper-converting machines, specially designed forming plates are used.

Cross folds are made by a blade which projects from a cylinder and either forces the paper between a pair of rollers, or presses it into a pair of gripper jaws in an opposing cylinder. The blade and roller system is faster but less accurate and is generally used for the cross fold in newspapers or similar products

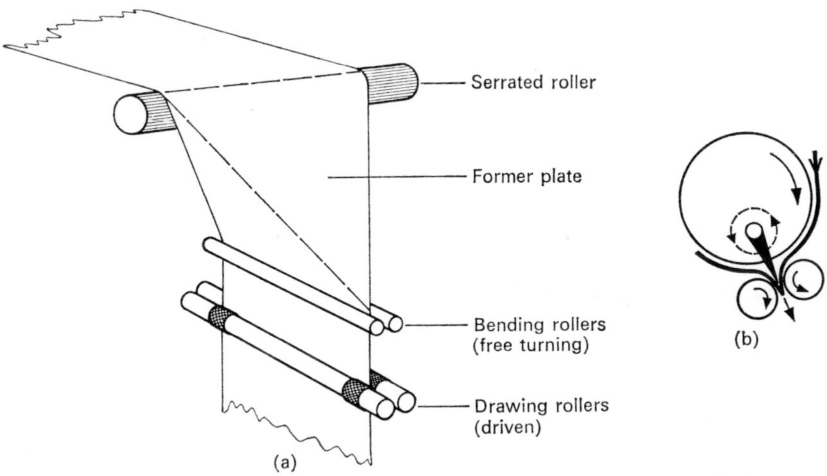

6.8 *Folding a web. (a) Principle of former fold. The web is pulled over a triangular plate and between a set of rollers to make a fold in the length direction. (b) Principle of blade and roller fold. The paper passes around a cylinder containing a blade which pushes the paper between a pair of rollers to form a fold across the web direction.*

where high speeds have priority over fine accuracy. For most other purposes the jaw-type cross fold is preferred for its more positive action, especially where more than one cross fold is required.

News folders. On many machines the unit referred to as the folder is really a collection of folding and ancillary devices coupled together to produce the desired end-product from the web. For instance, a typical news folder comprises: 1, a set of feed rollers to collect the various webs; 2, a triangular former to make the length fold down the webs; 3, another set of feed rollers (draw rollers) to pull the webs across the former; 4, a pair of cylinders which cut the webs to newspaper lengths and make the cross fold by pushing the sheets into 5, a pair of folding rollers; 6, the final delivery by means of a set of endless belts.

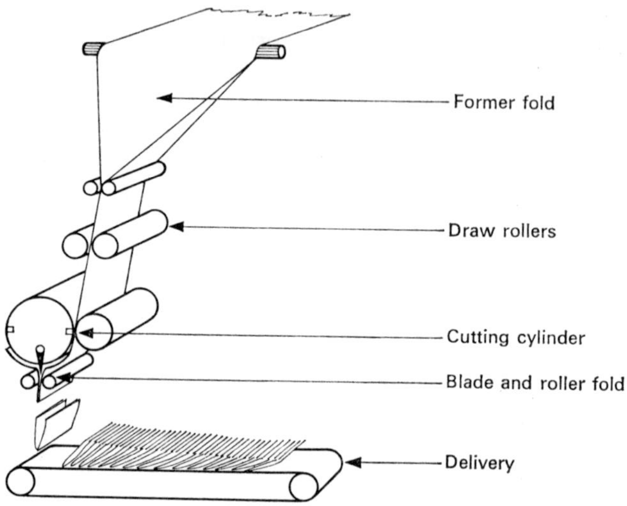

6.9 *Principle of a newspaper folder. A length fold is made over a forming plate before cutting to size. After cutting, a cross fold is made by the blade and roller method and the completed papers delivered on a set of endless straps.*

A news folder can be made more versatile by including a slitter on the first set of feed rollers so that the papers are produced half size (tabloid). The rest of the folder is hardly affected, but of course the format of the paper is radically changed. Another useful facility is the ability to 'cross associate', *eg* to bring a set of webs from one folder to another before the final cross fold so that more pages, or extra sections are obtained. A further variation which can be built into a news folder is 'collection' whereby the first set of pages on a two-set machine are diverted around a cylinder to mate with the second set, thus doubling the number of pages in the paper, although halving the number of papers per hour.

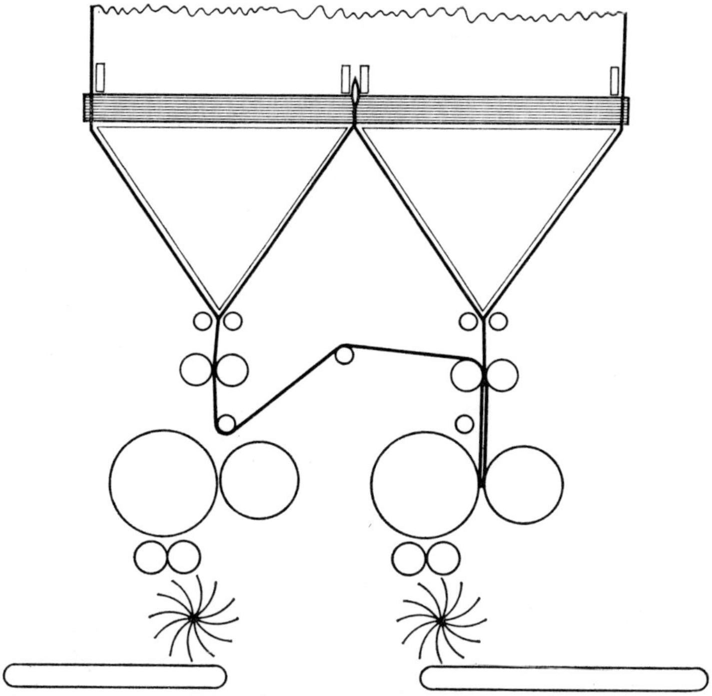

6.10 *Principle of cross association. The half-web from the left is associated with the other half-web after length folding but before cross folding. The result is a newspaper in two sections.*

Magazine folders have a completely different arrangement from news folders. They usually work from just one or two webs which are 1, slit into page-size ribbons; 2, diverted across angle bars; 3, registered to each other by passing each ribbon over an adjustable register roller; and 4, gathered together between a set of draw rollers. The combined strips are then given 5, one or two cross folds by the blade and jaw system contained in a set of cylinders. Full-size magazines will be given one cross fold and be complete, while 'digest' sizes will require a second cross fold. Delivery onto a set of endless belts is again the final stage in this type of folder.

Where two sections are to be collected on magazine folders, they will be

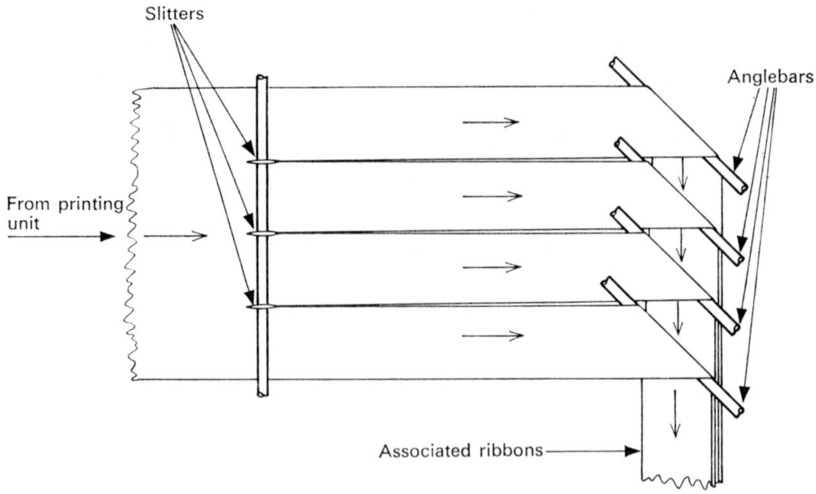

6.11 *Ribbon folding. For narrow-width publications, such as magazine and book sections, the web is slit into a number of strips after printing. The strips are associated for cutting and folding by passing each ribbon over an angle bar to align it with the others.*

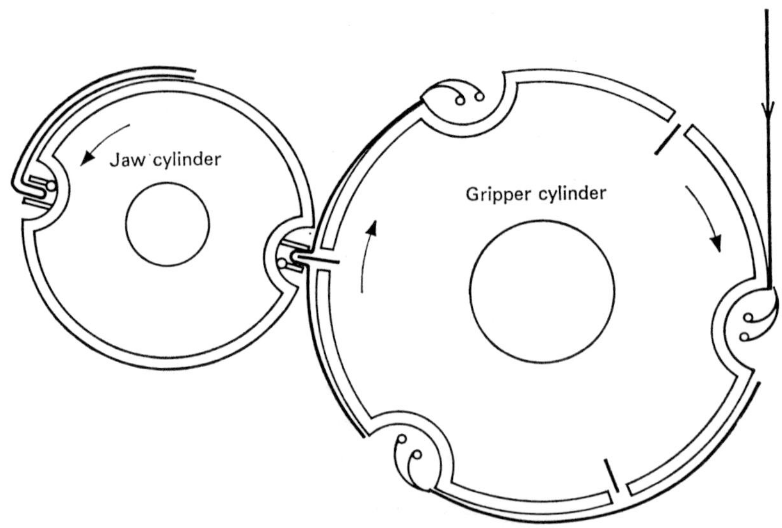

6.12 *Cross fold by the gripper and jaw method. This system is commonly employed for magazine production. Each revolution of the jaw cylinder delivers two copies.*

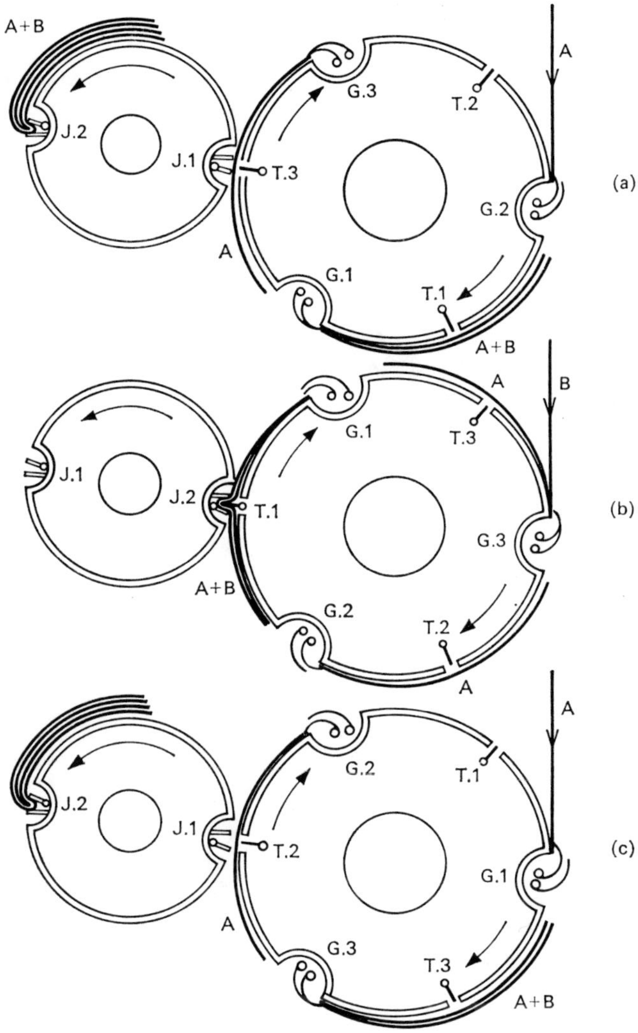

6.13 *Cylinder collection. Three diagrams showing the sequence of operations on a jaw and gripper folder whereby consecutive sections (A and B) in a publication can be collected together to increase the number of pages delivered. Each revolution of the jaw cylinder will deliver one (collected) copy.*

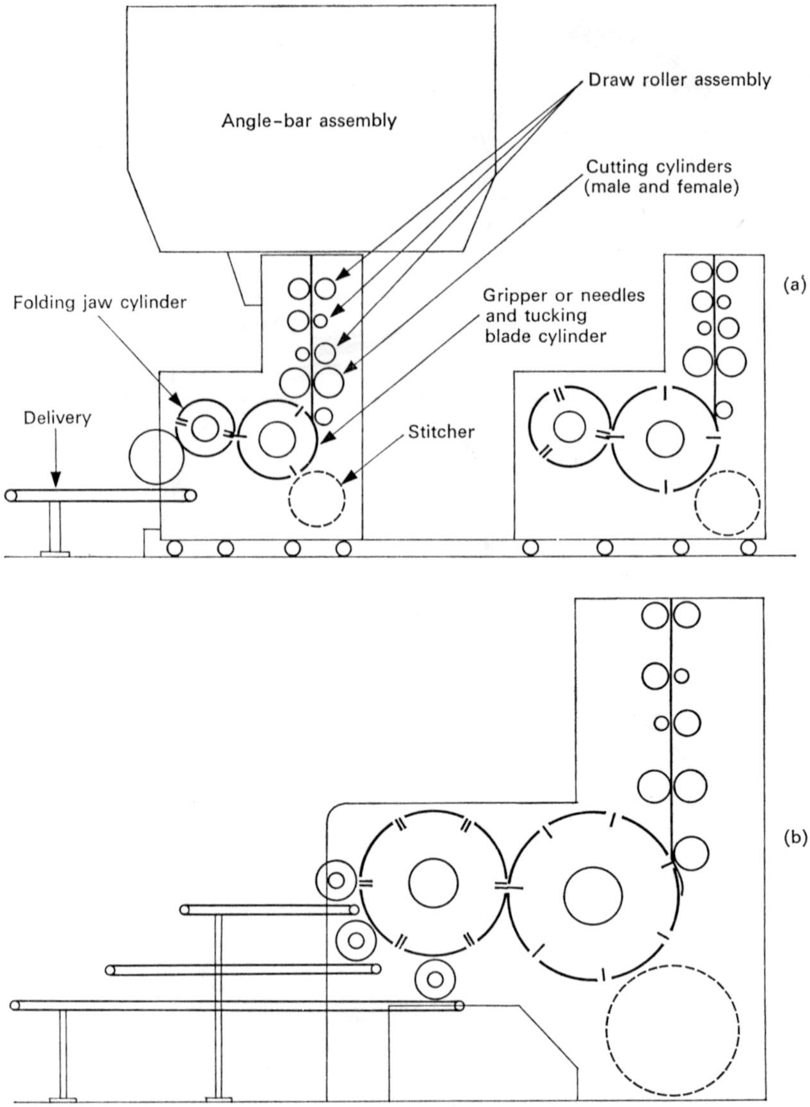

6.14 *Magazine folders. (a) A 3–2 folder for use with printing cylinders having four plates round is located to receive the assembled ribbons coming from the angle bars. This can be replaced with an alternative 4–3 assembly for cylinders having six pages around. (b) The 7–6 folder can be used with printing cylinders of four or six pages around.*

designed with an odd number of stations on the gripper cylinder and two sets of grippers at each station. A simple arrangement is to have a cylinder containing three gripper stations and tucking blades working in conjunction with a jaw cylinder having two sets of jaws (3–2 ratio). The tucking blades will operate every other revolution so that the second set of grippers at each station can collect alternate sections from the webs and forward the associated sections to the jaw cylinder.

It is considered to be an engineering advantage to employ gripper and jaw cylinders of as large diameters as practicable so that maximum time is allowed for each mechanical operation. Wear and tear on moving parts is reduced and sections are not unduly distorted when travelling around the cylinders. For this reason cylinder ratios of 5–4 may be preferred.

Where three sections are to be collected, the gripper cylinder must have at least four stations with three sets of grippers at each station, and a common arrangement is to have a 4–3 ratio folder. If a machine is designed to produce magazines of two collected sections at one time and three collected sections at another, it may be necessary to employ two alternative folders such as a 5–4 combination for collecting two-section work and a 4–3 combination for three-section magazines. A better arrangement in this instance is to employ a folder with seven stations on its gripper cylinder and six jaws on the opposing cylinder, which may collect either two or three sections, with the appropriate adjustments.

Variable size folders. A variable size folder is made to produce publications of different sizes from cylinders within a specific range. Apart from collecting, registering, cutting and folding with great accuracy at high speeds it is important to have means of adjusting the gripper hold on the copy and to advance, retard or twist a fold while the machine is running. It must be possible to disconnect the folder from the units in order to carry out maintenance and to register the folder to the printed webs. There must be a clutch system for opening the draw rollers in order to idle the machine without drawing paper and draw rollers should be adjustable to cater for publications of different size and thickness. The formers and angle bars must be adaptable for synchronizing the alignment of ribbons. A variable size folder must have means of altering the cut-off length by changing gears, or more generally by using positive infinitely-variable drives.

Zig-zag folders are used for the production of continuous stationery, where the work is required to be folded, but not cut. The blade and jaw system may be used, but with alternate folds made in opposite directions, causing the paper to fold up into a continuous 'concertina'. A different system for continuous stationery folding pushes the web towards a rotating coil, in which the paper becomes trapped into a fold. A pair of coils are required for these 'spiral' folders, to make the alternating zig-zag folds.

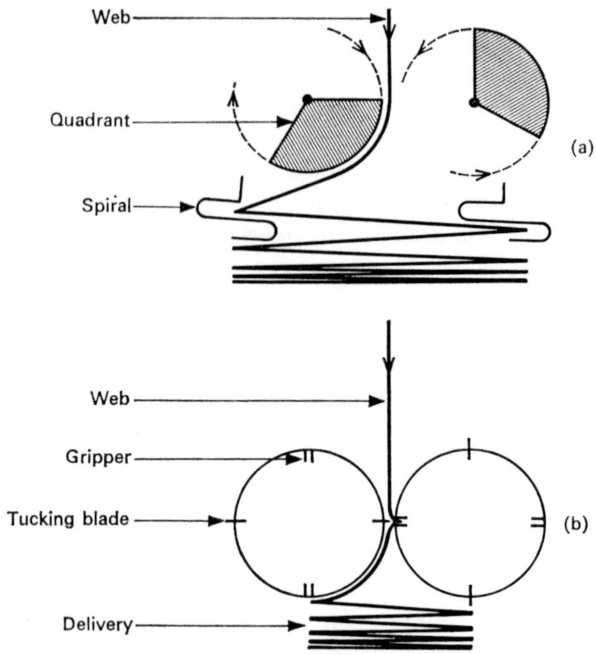

6.15 *Folding continuous forms. (a) principle of spiral folder. The quadrants deflect alternate forms towards the appropriate spirals. The partly-folded web is caught by the rotating spirals which complete the folds on the perforations. As the spirals turn, the folds move down on to the pack of work and the next perforation is captured. (b) Principle of nip and tuck folder. The tucking blades on one cylinder thrust the paper between the gripper jaws on the other. They close, making a sharp fold, and carry the paper a quarter of the way around the cylinder, where it is released. By this time the other cylinder has gripped it and makes a fold in the opposite direction.*

Rewinds

Where the printed work must have further processing, it may be rewound at the end of the printing machine. Rewinds work in two basic ways, one by friction drive on the surface of the reel, and the other by driving through the centre of the reel core.

Surface drives are generally simpler, and produce a firm, tight rewind similar to that of the mill reel. The most elementary system is a drum geared directly to the machine drive, with the reel being pressed on to it by means of weights or springs to ensure traction. Web tension may be ensured by gearing the rewind cylinder to travel 1–2 per cent faster than the surface speed of the printing cylinders. Whereas the reel brake exercises a retarding force on the reel to maintain web tension before printing, the rewind exerts an accelerating force to achieve a similar web condition after printing.

A two-drum system is often preferred for better rewinding efficiency, as the reel is trapped in the gap between the two cylinders, and there is consequently less need to employ weights to ensure traction. For complete control the rewind may be driven through a positive infinitely-variable gearing system, so that the amount of tension may be accurately kept under changing conditions.

Centre drives. The frictional force which controls the surface rewind can also mark the wet print and will be less efficient with slippery, non-absorbent or extensible materials such as plastic films. For this sort of work a centre drive is preferred. This type of rewind must have variable-speed control as the reel core must turn progressively slower while the rewind becomes larger. However, although the rate of rotation decreases, the power required will increase as the reel becomes heavier. Therefore the centre drive rewind must provide for variable speed with constant torque to retain a constant web tension.

Automatic rewind changers. Where high speed production of long run work is undertaken it will be an advantage to have an automatic rewind changer. This device is based on the principle of the revolving reelstand as described under automatic reelchangers, and its method of operation has several similarities. The reelstand is revolved as the rewind is nearing its maximum size, so that the glued reel core approaches the web. The core is accelerated to web speed and contact made with the web, at which time the full reel is cut from the web.

Perfecting a web

To print both sides of a web requires two printing couples. These can be either completely separate printing units or combined to make a composite perfecting unit. Where separate couples are employed there is usually more time available after printing the first side for ink penetration and setting. The designer also has more flexibility to position the units according to the space available. This may be especially important where further colour printing units are required on the machine, or where a machine must fit into a restricted space in a production line.

Most publication machines employ the arch-type perfecting units for text printing. With this system the two plate cylinders, impression cylinders and inking systems are arranged in a compact arch shape so that all working parts are readily accessible for plating-up, cylinder dressing and roller setting. If the cylinders are correctly packed, web control is particularly good in this type of unit due to the paper wrapping around the impression cylinders.

A disadvantage of the arch unit is that the second impression follows so quickly after the first. On a news machine running at a web speed of 10 metres per second the second print may follow less than one tenth of a second after the first. Fortunately for newspaper printers the combination of absorbent newsprint and the thin, penetrating news ink gives a very rapid method of drying. Nevertheless second-impression set-off can occur if heavy ink films are used or the ink is too thick, particularly at high speeds. For this reason it is especially important to clothe the second impression cylinder with a blanket having an ink-repelling surface (see Chapter 4).

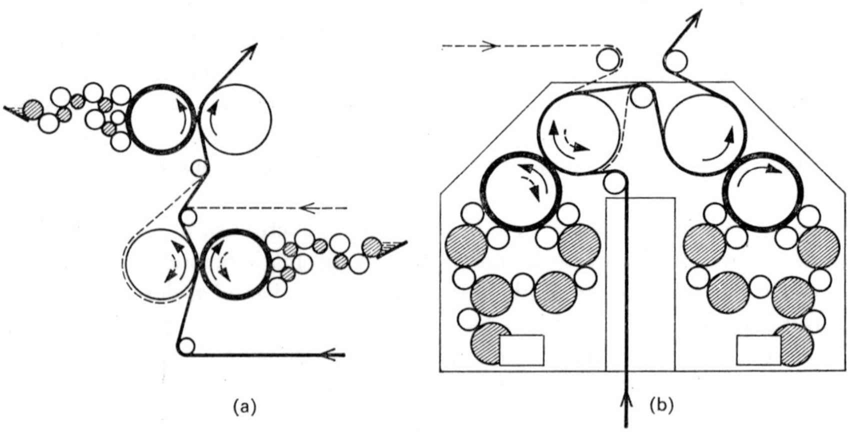

6.16 *Versatility of web printing. (a) Two separate printing couples, showing a perfecting lead with the unbroken web line and an alternative lead for two-colour printing indicated by the broken line. The rotation of the first couple must be reversed when the web lead is changed. (b) Perfecting and colour leads are also possible on an arch-type unit provided one couple is reversible.*

Multicolour web printing

If several colours are required on a web they must all be printed at one pass through the machine. Due to web control problems in a roll of paper which may be several thousand metres in length, it is not generally a practicable proposition to pass a web several times through a machine to print multicolour work, although this may be a satisfactory procedure for sheet work.

The simplest method of producing extra colour is to add to the perfecting unit another plate cylinder with its own inking system, but printing against a common impression cylinder with the text. This arrangement is quite suitable for an odd spot of colour, or where there is no overprinting, but for comprehensive colour printing additional complete units must be provided. These can be couples arranged above the text units to conserve space and shorten web leads, or they may be specially-designed four-colour units if the machine is intended to undertake colour halftone work.

Reversible units. It may be desirable to design printing couples for additional colours so that their direction of rotation is reversible. This gives the printer 'colour flexibility' as the unit may print on either side of a web by bringing the paper into the unit from the opposite direction when the cylinders are mechanically reversed. When this type of unit is designed it is important to ensure that its performance is equally satisfactory in either direction for basic functions such as ink distribution, forme rolling, impression, web control and plating-up. It would obviously be difficult with a lithographic system as the printing cycle would be reversed.

Non-publication work such as packaging or stationery printing are not so much dominated by the requirements of text printing. The nature of the end-product will greatly influence the arrangement of the printing units. For instance, a machine for printing on a fairly inflexible material such as boxboard will have its units built in-line so that the board can take a fairly straight path through the machine.

For printing on more flexible materials the units can be 'stacked' on top of each other to save space, and colour flexibility obtained by making some or all units reversible. Stretching materials such as plastic films, will be printed best on a machine where all the plate cylinders are grouped around one impression cylinder to ensure that the register does not vary.

The common-impression cylinder principle is also used for producing colour magazines where halftone colour may be required on any or every page. For rigidity, the impression cylinder is usually quite large, and five or six plate cylinders are often grouped around it so that 'special' colours can be printed as well as the colour halftones. If the reverse side of the web is also required to be multicolour printed then a second unit of the same type will be used.

Any pressure adjustment on these machines will necessarily have to be done on the plate cylinder. Attention must also be given to correct ink trapping by controlling ink tack, and to the limiting of ink film thickness by employing undercolour removal during platemaking (see wet-on-wet printing in Chapter 3).

Before backing-up the first side the ink must be perfectly dry so that marking up under pressure on the second impression cylinder does not occur. For this reason the web passes through an oven dryer before travelling to the second unit. After perfecting, the web passes through the dryer a second time to ensure that the ink is dry before entering the folder.

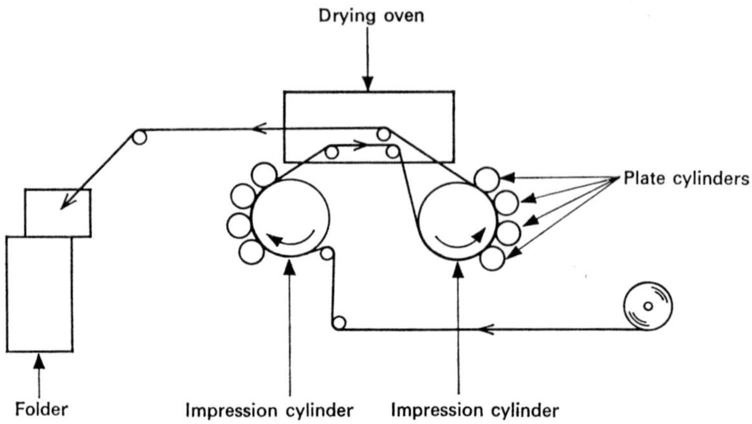

6.17 *Web-fed common-impression machine. This type of machine is used for magazine or catalogue work where multicolour printing is required on many pages. Two units are required to perfect the web. Inking systems may be arranged to move away from the plate cylinders for accessibility.*

Multi-web machines

Several webs may be used for large-scale production of newspapers, magazines, books, stationery and packaging materials. For efficient working an overall control system should be built in to the machine to co-ordinate the functions of the various webs, printing units and other operations which may be undertaken at the same time.

Each aspect of control should be governed from a centralised position to avoid faults in any one area holding up the whole production line. A comprehensive central control console could incorporate provisions for remote operation of reel brakes, sidelays, feed rollers, impression adjustment, ink metering, inter-unit registration, web-to-web register, insetting, drying system and folder adjustments, as well as the usual press controls such as starting, stopping, inching and speed variation.

Multi-web machines may also incorporate separate sections utilising different printing processes. Offset colour sections from a common-impression system produced lithographically may be fed into letterpress news sections; similarly photogravure colour may be fed into flexographic book production and so on. Such a machine becomes a multi-process production line capable of producing a variety of products.

Variable-size web machines

It is a common misconception that web printing entails restriction to one size of printed product. In fact there are several ways in which different print-lengths may be produced on a web machine.

Cylinder division is the simplest system, whereby a cylinder size is chosen which will allow a number of suitable divisions for a range of useful print-lengths; *eg* a cylinder with a 960 mm circumference may be used to print jobs with a 960 mm print-length, one up, or 480 mm jobs two up. This cylinder would also be suitable for any jobs with print-lengths which were multiples of 3, 4 or 5 mm, which were also factors of 960 mm. A change in printing process does not affect the validity of this method.

Change-size machines can be used where the above system is not suitable. The machines may be designed either to change complete printing units having different dimensions, or more simply to change only the printing cylinder.

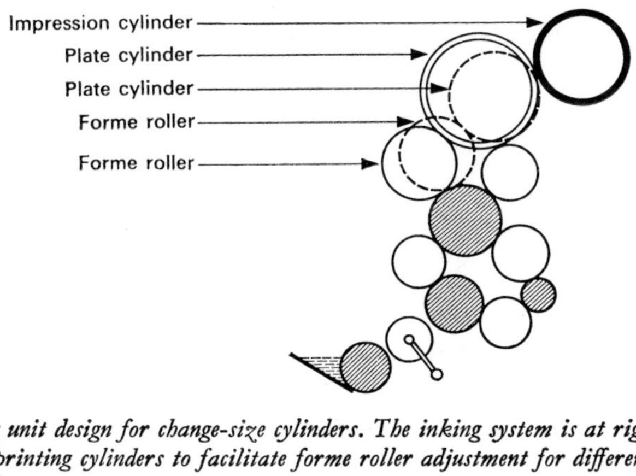

6.18 *Printing unit design for change-size cylinders. The inking system is at right-angles to the printing cylinders to facilitate forme roller adjustment for different sizes of cylinders.*

This second method is not practicable for offset systems unless plate and blanket cylinders are changed together.

Intermittent web-feed machines are more versatile still. On this type of machine a variable web feeding device is used which can be set to meter out any length of paper required up to the length equivalent to the plate cylinder circumference. If a shorter length is metered out, the paper feed is stopped for the period while the cylinders finish their revolution to the start position.

This system is understandably slower than constant-feed machines, but the versatility, especially for short-run work, makes it a very useful method for producing an infinite range of sizes on one machine, without the need for a large stock of spare parts.

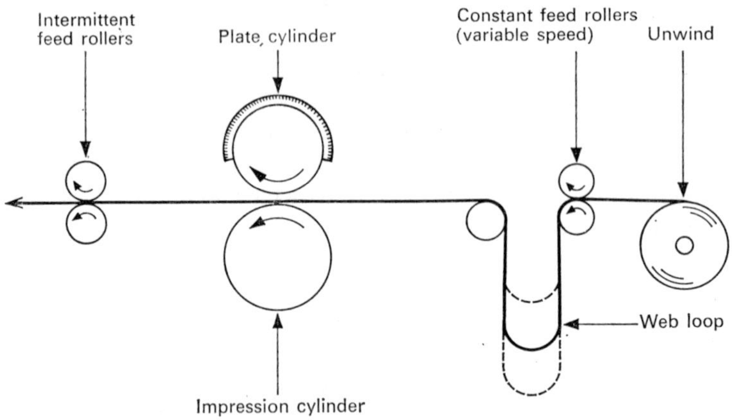

6.19 *Variable print length via intermittent web feed. The intermittent feed rollers can be set to pull through any required length of paper at each revolution of the printing cylinders. The web stops when the appropriate length of paper has been printed, but the cylinders keep turning until they are again at the start position, when the feed rollers again begin to pull the web. To avoid snatching the reel, a constant feed roller station is located at the unwind. The speed of these rollers is variable so that their average feeding rate is the same as that of the intermittent feed. The loop of paper acts as a reservoir to allow for the periodic differences between the two feeding stations.*

Reel-fed flatbeds can utilise the intermittent system to feed flatbed machines with reels. The reciprocating movement of a cylinder or platen machine can be made to coincide with the intermittent feed of the web, and the variable-feed device used to obtain different print lengths.

Constant unwind. Control of reel unwind requires special attention on intermittent feed machines as it is bad practice to snatch at the reel by rapid starting and stopping. This difficulty is avoided by providing a separate feed for the reel to unwind it continuously, but at a variable rate to match the *average* of the intermittent feed. Between the feeding unit and the printing units there will be a free loop or festoon of paper which acts as a reservoir to accommodate the differences between the intermittent feed through the units and the constant feed of the reel. It can be seen that with this system, web tension control becomes independent of the reel brake, and is only controlled by correct feeding between the printing units and other operations.

WEB OFFSET

Print requirements

The higher speed of lithographic web offset minimises damping as a problem, since far less water is needed. For this reason, the reservoir capacity of the damping mechanism need not be high, and stockinette sleeves are suitable covering for conventional damper systems. Usually only one plate damper is used and systems requiring no covering at all to the plate damper are frequently favoured. Usually uncovered dampers rely on alcohol enrichment to reduce the necessary water output and apply moisture either directly to the printing plate or to the first plate inker.

Since printing is continuous the fountain supply also should be continuous rather than oscillatory. Variable speed feeds are therefore used, but start-up needs to be carefully planned to get the right water balance quickly to avoid waste in production. This can be effected by hand-operated sprays, by a machine flood button to apply excess on start-up, or by console control of an electronic digital logic system of pre-damping. Care must be exercised not to overdamp, since transfer of water to the web introduces a likelihood of a web-break at that point.

Inking systems for web offset presses can be conventional in design but need to be continuous in supply and satisfy special production requirements. Continuity is effected by replacing the conventional reciprocating ductor with an ink transfer roller. This may be a segmented roller allowing adjacent segments of the roller to be spirally eccentric to give a haphazard contact line to both ductor and first roller in the ink train. Brush feed can be used allowing lighter ductor blade settings and thus reducing friction at the ductor. This may be important with heatset inks containing hydrocarbon oil fractions where it is not desired to set the boiling point too high. The inking system may be designed to include water cooling through the steel distributors to maintain the ink film on the rollers at ambient temperatures.

All types of litho plates are run on web offset presses but the type of run normally demands a smooth surface for easy damping, good selectivity between image and non-image areas, and toughness of the image for long runs at high speeds. Additionally it is desirable that the metal structure should easily accommodate the bend at the plate clamping gutter without hump formation to cause roller bounce. Plate mounting and lockup are simplified to a single preformed bend at each end of the plate and quick locking clamps operated by a key or spanner.

On wider cylinders, plates may be staggered so that the gutter does not continue in the same line across the complete cylinder. Unless the press is designed to accommodate both litho and letterpress plates, cylinders may be cut to take thinner caliper plates in the 0·3 mm range. With deeper cut cylinders, underpacking sheets allow thin caliper plates to be used as required. Bi-metal plates most readily satisfy web offset requirements. Aluminium plates stand up well to press wear but are prone to crack at the gutters if used in thin calipers owing to the tensile build-up under compression. Metal should have its grain in the same direction as the web to minimise this hazard.

Open or in-line presses are generally custom-built machines of unit construction similar to sheetfed press design. Printing is on one side only, and it is necessary to use a web turnover section in order to perfect. Such machines are utilised to print any product normally produced on sheetfed machines. Personalised cheque printing using magnetic inks can be accommodated on the same machine as relief plates printing moisture-set security ink, by employing inter-unit dryers. Carton board can be fed from reels to offset printing and inline cutting and creasing units with an inline sheeter to allow conventional stacking. Continuous stationery may also be printed on such presses, linked with finishing and collating.

Blanket-to-blanket unit perfecting presses constitute the most widely used web offset system. A vertical web lead is preferred for newspaper production for reasons of access already discussed. The unit can be designed with inkers supplying the printing plates from above the central cylinder position and damping arranged below the plate cylinder. Alternatively plate and blanket cylinders can be arranged at the top, with both inking and damping below the unit. With this method the damping unit is less accessible, but the arch arrangement allows for the addition of an extra plate cylinder to print 'spot colour' on the same blanket on one side of the web without any loss of paging. Heaters are not normally employed on this type of press.

Machines for good class heatset colour work more frequently employ a horizontal web lead. Single level access to all colour units is a good working arrangement. It is possible to limit sheet marking from idler rollers using the horizontal plane layout, and an in-line dryer can be accommodated in such a grouping. Units can be designed with interchangeable cylinder assemblies to provide variable cutoff. Change of format is rapid and consists of wheeling in the interchangeable assembly, locking it into electrical and lubrication connections hydraulically. Facilities are available to print any combination of single and multi-colour webs as the machine has unit assemblies.

Where four units are used for colour work, changeover to black and white copy may lengthen makeready times. Additional units can be used to print single colour copy which can be mated with a colour spread if desired, with or without oven drying. The additional units also give the ability to prestretch

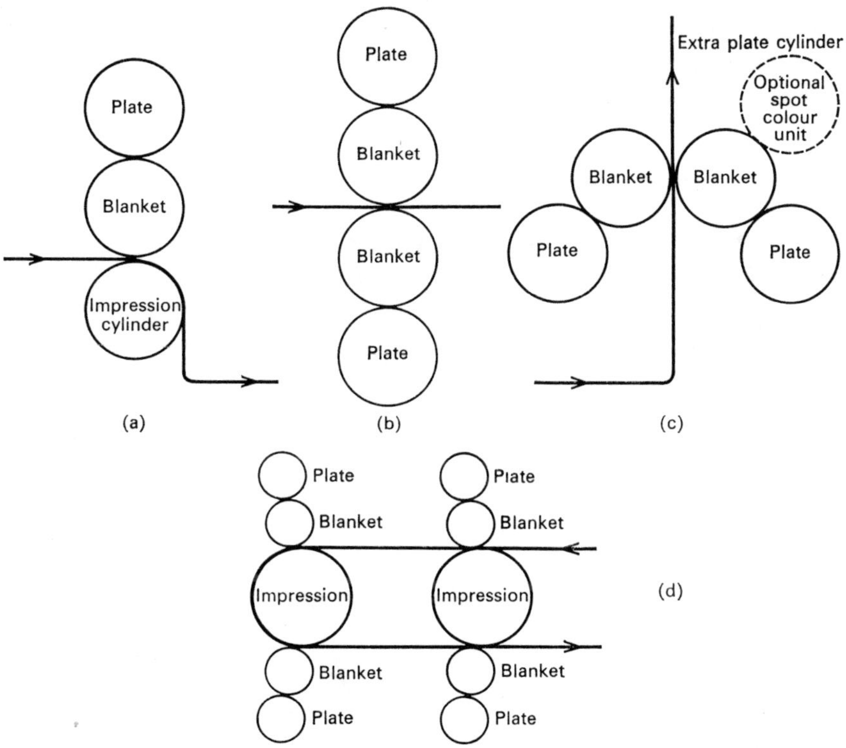

6.20 *Web offset arrangements. (a) The open or in-line press. (b) Blanket-to-blanket, horizontal web lead. (c) Blanket-to-blanket, vertical web lead. (d) Double cylinder system.*

the paper in a blank press pass before the first printing unit, which helps register for colour printing and removes fluff from the paper surface which would otherwise detract from print quality. Preconditioning units may be included in the web feed to standardise web input.

Satellite presses. A third type of web offset machine is that built on the common impression principle. The required number of printing units may be built round a large drum that serves as a common impression cylinder to form a satellite system with its surrounding blanket cylinders. Alternatively impression may be split between a double cylinder or twin-satellite system providing shared impression in any number of combinations.

Single impression is advantageous to register control as the web is carried over a large surface cylinder area. The web must be dried and turned over for perfecting and must be reconditioned before the second pass to prevent register variations between the two printings. The use of satellite systems linked in with unit perfecting systems enables monochrome to be run on the unit press whilst the satellite is being made ready for colour.

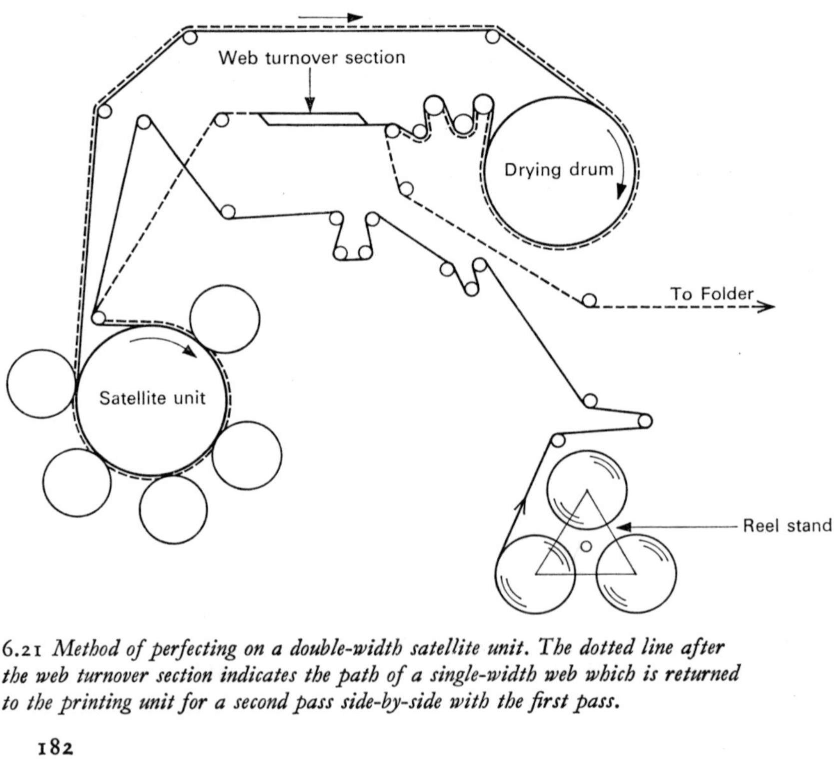

6.21 *Method of perfecting on a double-width satellite unit. The dotted line after the web turnover section indicates the path of a single-width web which is returned to the printing unit for a second pass side-by-side with the first pass.*

Web offset problems

Good quality lithographic printing on the web is only possible with good copy preparation. Fine hairlines can be lost with high-speed lithography and should be avoided in the copy. On the other hand the machine system has a direct bearing on tonal gradation since a blanket-to-blanket press will print a softer dot than a satellite under identical conditions and dot gain in the highlights will lessen contrast.

Absorbency is a characteristic required in newsprint, but the lower film-weight achieved by offsetting modifies this need to some extent. Nevertheless litho inks must be absorbed sufficiently rapidly to prevent marking and set-off. As the web is run with the grain in the machine direction and low contact time in the impression nip with low tack inks and minimum press moisture, weaker bodied papers can be run than for sheetfed work. Surface treatment with starch and other chemicals has a limited use because it is film-forming and retards ink penetration. The desired print gloss and density are closely associated with smooth, bright stocks of low absorption characteristics, and these properties are usually enhanced by including a higher proportion of china clay in the furnish or by supercalendering. Such qualities in the stock increase setoff, and introduce a mineral content in the linting which occurs with open-structured web paper. A build-up of fibre from the paper furnish can usually be tolerated as it does not interfere greatly with printing unless a press stoppage occurs. Conversely a build-up of mineral pigment matter on the blanket accelerates plate wear and becomes difficult to remove from the blanket.

Non-heatset papers for web offset are two-sided in respect to linting, the felt side being worse than the wire side. On blanket-to-blanket presses, the paper emerges from the impression nip hugging one blanket more than the other. More linting occurs on the latter because there is a greater angle between blanket and web, and ideally the web should be run with the wire side to the more demanding blanket. The use of serrated rollers between units is intended to balance this tendency to hug one blanket. Some machine manufacturers employ a staggered cylinder arrangement to give more wrap at impression to apply proper tension without utilising serrated rollers which may smear the print.

Use of bleached pulps and careful control of calendering and sizing minimises the requirement for fillers but provides a compromise between setoff and print quality. Linting can be reduced by running more water, providing inks used have sufficient internal cohesion to prevent them becoming waterlogged.

Paper for heatset lithographic printing can be made to the right opacity and surface-sized to prevent linting. Coated surfaces provide absorbency to lithographic inks coupled with a smooth finish. Problems arise through blistering due to the closed coated and printed surface lifting from the paper when moisture vapour cannot leave the stock at high oven temperatures. Flame temperature and machine speed, coupled with the porosity, strength and moisture content of the paper are controlling factors. The amount of moisture in the stock needs to be specified regardless of paper substance since the degree of evaporation is not dependent upon paper caliper.

Delamination of the sheet through lack of fibre bonding in the body paper is a more likely mechanical failure than surface pick because the low tack printing conditions in the nip are less hazardous than the mechanical stress of high speed impression. Uniform ink receptivity is the important factor and this is influenced primarily by the binder constituent. Delamination may be caused by the effect of ink tack simultaneously on both faces of the web rupturing the stock.

Since impression is so important for consistent web feed through offset presses, damage to blanket surfaces must be avoided. Mill joins should be kept to a minimum and clearly marked so that the machine crew can release pressure contact as they pass through the press. When this cannot be done the press should be slowed for the passage of the mill join to give time for any necessary emergency action to be taken to preserve production flow. Packing gauges help in enabling blanket surfaces to be checked under machine tension for evenness and uniformity of pressure. Low spots may show up under cylinder tensioning which are not apparent off the machine.

Colour density is a problem with lithography, and target ink levels should be fixed before running, so that the press can be controlled within these tolerances. Inks should be chosen which run at a reasonable film-weight without setting off. Typical inks may have a modified phenolic base of fairly high plastic viscosity with thixotropy induced by suitable catalyst additives, allowing good wet-on-wet trapping and excellent holdout for heatset colour work.

If a satisfactory ink film can be carried without ink spread occurring, better gloss and intensity is possible than on sheetfed printing since the faster drying retains specular reflective characteristics in the dried ink film. Solvent additions depend upon the type of work being printed. Non-drying mineral oils are used for newsprint where drying is primarily by penetration, but for smoother and less porous periodical or book papers the use of quickest resin inks requires a drying oil addition. For heat-set inks the hydrocarbon oil fraction must evaporate at oven temperature but not on the press. In this case solvents are hazardous to normal blankets and rubbers must be chosen that have good resistance to ink solvents.

Print and register conditions are similar to those for multicolour sheetfed work. Register can largely be controlled by the web once cylinder makeready has been completed. Apart from circumferential and lateral adjustment of cylinders, helical adjustment is possible by angling the web or by imposing a slight twist of the contact point between cylinders. A swing of up to 0·5 mm in either direction is possible on some machines. Some contact pressure is lost when cylinders are out of parallel and a reduction in overall print quality may occur from such adjustment. Accurate platemaking will avoid the necessity for plate slewing. Plate bending must be positive; on larger size work hydraulic presses are used for pre-bending plates to conform to the house system of registration and this may link in with pin-register assembly at the platemaking stage and use of pin-register on the press.

ACCELERATED DRYING SYSTEMS

Chapter 3 dealt with the principles of ink setting and drying. Drying of ink on reel-fed machines requires a different basic approach to sheet-fed machines, and the following factors should always be taken into consideration.

1 Substrate. The principal property affecting drying is absorbency, others being the moisture content of porous stocks and the effects of atmospheric changes on the material.

2 Ink. The most significant factor is the drying time at normal room conditions of temperature and humidity: other factors include ink film thickness, additives, solvent evaporation rate and boiling range, and the effects of atmospheric changes.

3 Machine design. An important feature is the time interval between one printing nip and the next, *ie* the length of web between the nips divided by the speed of the web. Other design features which affect drying are the path of the web, ancillary operations and provision of equipment for accelerating ink drying.

The fundamental difference between setting and drying must be emphasised. If an ink is set it can be overprinted, passed over idler rollers and taken through certain in-line operations, but until it is dry it should not be rewound, folded, laminated, stacked in high piles or otherwise heavily worked.

Drying systems are built into machines in order to obtain higher speeds. This is done by accelerating two basic drying functions: 1, the removal of unwanted phases in the ink; and 2, changing the state of ink from a liquid to a plastic or solid. The mechanisms of drying systems are therefore concentrated on control of air movement and temperature control. Air movement is particularly significant where volatile solvents are being given off; passing a stream of air over the web will accelerate evaporation, and removal of solvent-saturated air is essential to permit rapid evaporation. Temperature control is important as evaporation occurs more rapidly at higher temperatures due to increased molecular activity, and where there is also a degree of oxidation drying or polymerisation, this will also be accelerated by heat. Some thermo-setting systems dry by thermal catalytic action, without the presence of solvent, induced by high energy radiation.

Some of the basic items of equipment used in drying systems are 1, air control by blowers, extractors and circulation systems; 2, temperature control by electrical heating elements, steam-heated drums, gas flame impingement and microwave radiation. Infra-red or ultra-violet radiation may also be used in special circumstances, mainly on slow-running machines.

Each of the above items may be used on its own in specific instances, but a comprehensive system would include a combination of elements. The complete system will be either 1, an integrated drying system comprising different elements at various places in the web path; 2, a drying hood, as used commonly with gravure printing units; 3, an oven or tunnel.

The choice of method will also be governed by the requirements of safety

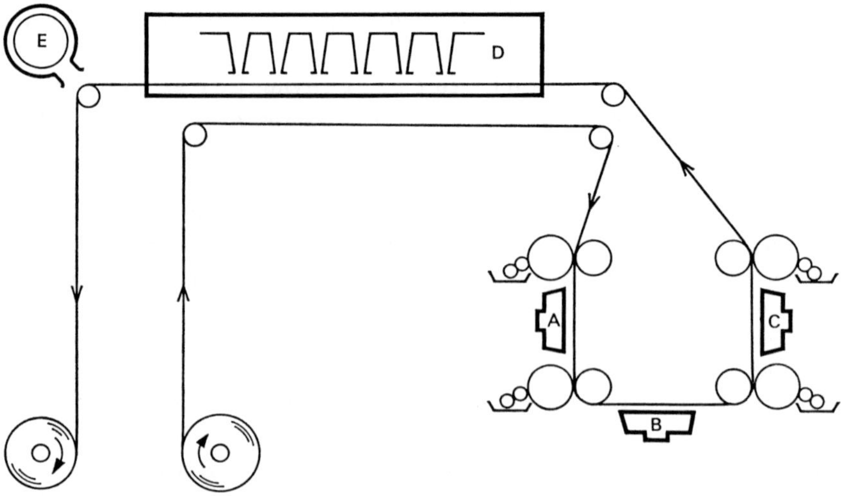

6.22a *Integrated drying system on a flexographic machine. A, B and C are hot air blowers between printing units to set the ink sufficiently for overprinting. D is the main drying tunnel containing heating elements, air knives and an air circulating system for removing the solvent phase from the ink. E is a cold air blower for cooling and setting the residual ink prior to rewinding.*

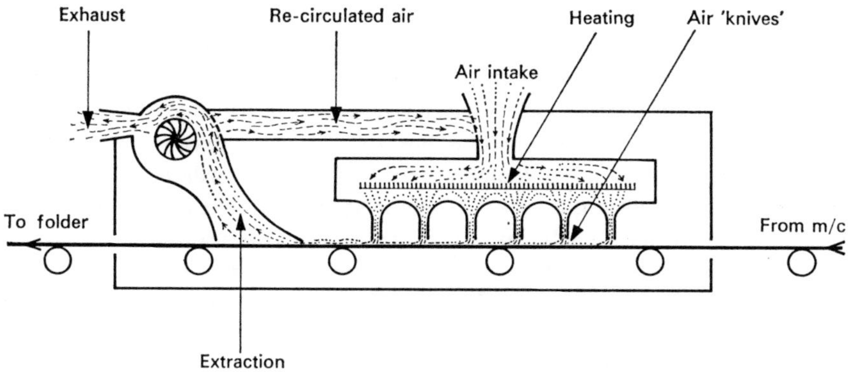

6.22b *Principle of hot-air drying system. Air coming into the dryer is first heated and then forced through slots to impinge on the web at high velocity and penetrate the dead layer of air adhering to the web. This 'scrubs' the web and accellerates the evaporation of solvents. Solvent-laden air is then extracted and either exhausted or re-circulated. The degree of exhaust or re-circulation may be controlled by adjustable vents or flaps.*

and the degree of contamination by exhaust. For this reason the use of direct flame and internal electric systems are ruled out where flammable gravure and flexographic solvents are present. Rapid exhaustion of volatile fumes must be considered in relation to pollution and solvent recovery systems may be essential. Resins contained in heatset inks for litho or letterpress usage are basically thermoplastic and soften on the application of heat to permit penetration of the stock and evaporation of the solvent. For multi-colour offset printing dryers must be efficient and capable of dealing with a film density of 300 per cent at specified press speeds. The thermoplastic nature of the ink demands efficient cooling before a scuff-proof ink film can be obtained. Infra-red units are particularly useful for interdeck drying or for setting moisture-bound inks. Microwave dryers are specifically suited to polymerising ink systems.

Hood dryers

On gravure machines the web is usually taken to a hood directly above the printing unit so that drying commences as quickly as possible after impression. Each printing unit will usually have its own dryer and the web is taken through the hood by passing over a drum or a bank of idler rollers. Where drums are used they may be heated by electricity or steam while other systems employ hot air impingement and extraction.

It is desirable to keep web leads as short as possible to economise on materials and control register; some hoods allow the web to be taken out from a slot at the top, directly to the next unit, where extended drying is not necessary. Where possible it is advisable to avoid heat to prevent embrittlement and distortion of the substrate, but for high-speed drying heat is sometimes necessary. Operating temperatures will vary according to materials, speeds and dryer design, but typical temperatures for plastic films at normal speeds would be 60° to 75°C and for papers and boards, around 75° to 85°C.

Where air impingement is used, 'air knife' slots or nozzles direct high velocity air at the web. This is coupled with an extractor system so that once the web is 'scrubbed' with air and the solvent vapours lifted from it the saturated air can be quickly removed. Although hot air is more efficient to accelerate evaporation, cold air may be preferred for removing the solvent vapours as this may help to reduce web temperature and set the residual ink film. The efficiency of this type of system depends on many factors such as the number of nozzles or knives, their distance from the web, length across the hood, velocity of the air and its rate of extraction. An efficient system should enable web speeds up to 7 metres per second to be achieved.

Hoods may be designed on the recirculating air principle whereby adjustable baffles or dampers can control the amount of air that is exhausted while the remainder is re-circulated. This enables the quantity of solvent-laden air in the hood to be accurately controlled and also ensures sufficient saturation in the exhaust system to provide for efficient solvent recovery.

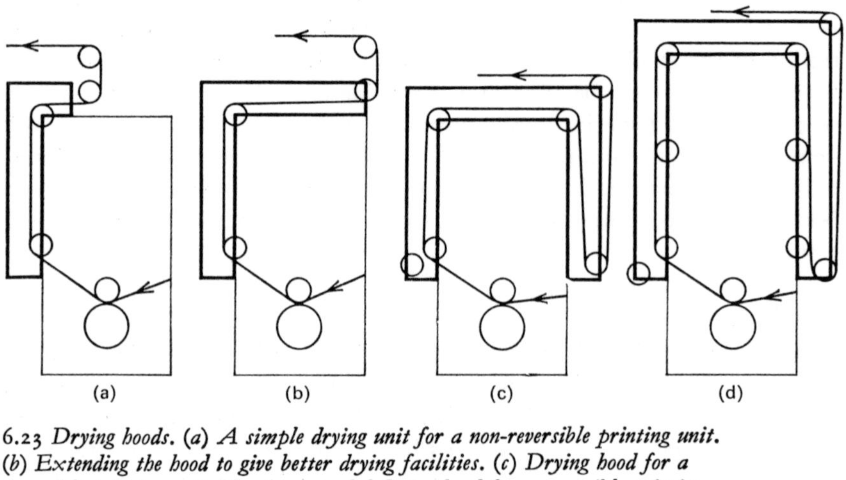

6.23 *Drying hoods. (a) A simple drying unit for a non-reversible printing unit. (b) Extending the hood to give better drying facilities. (c) Drying hood for a reversible printing unit. (d) An extended drying hood for a reversible printing unit.*

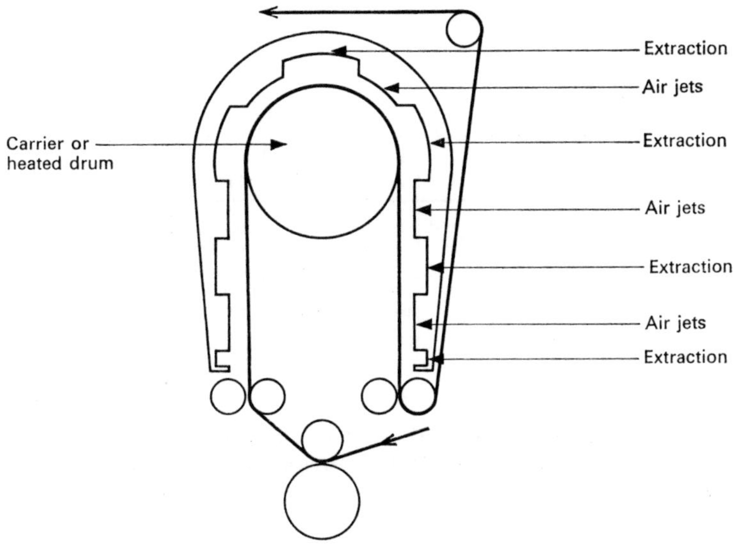

6.24 *Drying hood with alternate application and extraction of air. After printing, the volatile phases of an ink can be removed by air turbulence which carries away the unwanted solvent content.*

Gas heated drying ovens

Heatsetting letterpress and web offset inks require high temperature drying since solvents have to be evaporated and polymerisation accelerated in a resin-curing system. Suitable drying ovens perform three basic functions: 1, direct impingement from a gas flame vaporises the solvents; 2, heated air maintains a controlled temperature to dry the resins; and 3, the web is reconditioned by cooling and possibly raising moisture content.

Obviously at slow speed a direct flame will ignite a web of paper. At higher speeds however the web becomes insulated by a cushion of air which it carries into the oven and cannot be dispersed by the direct flame in the short period of contact with the web. Facility is therefore provided to cut the gas supply to the burners if the press is stopped or slowed, or to move the direct heat away from the web. Gas from the town supply or bottled gas may be used to fire the oven and has the advantage of cleaner working than fuel oils which are cheaper but burn with a sooty flame. According to the oven used, burners may vary in flame conditions; controlled pressure or combustion burners may be used in which case there will be a number of them spaced across the oven, or a 'lazy flame' will be used which is a larger flame of lower

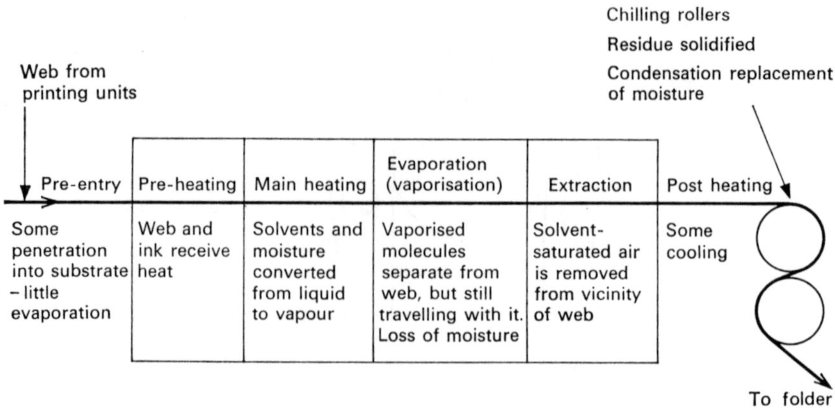

6.25 *Principles of heat-drying systems. The volatile fractions of a heat-set ink must be removed before drying can be said to be completed. The diagram shows the phases through which an ink film must pass before complete drying occurs.*

pressure that tends to be carried along with the web and insulated from it by its carrier film of air.

A super-heated air drying section follows the direct flame and is quite an efficient drying unit on its own. However, it is considered slow to respond to abrupt changes in web speed and the direct flame overcomes this deficiency

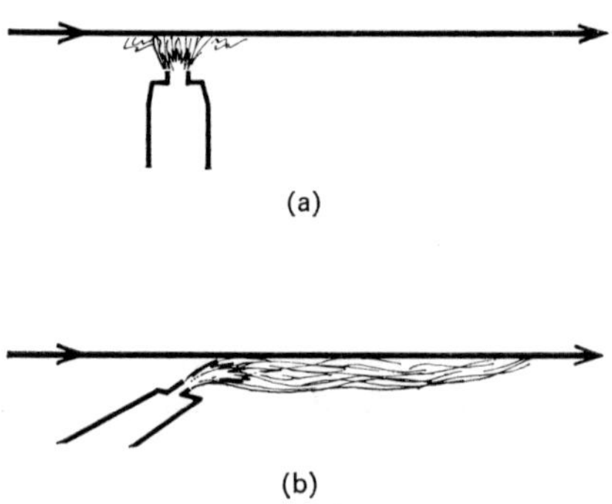

6.26 *Heat setting by gas flame—penetrating the 'dead layer'. To cut through the close layer of cool, non-conducting air which adheres to the web even in a drying oven, two types of flame are used. (a) The hard flame, which is directed to the web at high velocity to cut through the dead layer; many jets at high temperatures are commonly used. (b) The soft flame (lazy flame) is directed at low velocity to the web, travelling along with it to penetrate the dead layer. Fewer, but larger jets are required with this system.*

by cutting in at the required speed which is usually somewhere between one and three metres per second. Heat from the gas flame cannot be dissipated quickly: the air knives need to be linked therefore to a thermostat control to retain constant web temperature. Temperature recording charts should be part of the oven control equipment and enables a post-reference to be made at any time as a check against a production run.

Controlled air turbulence breaks down the stagnant heat layer so that the web can be cooled and fumes exhausted from the oven. Chill rolls are normally used to cool the web at the exit section of the dryer. Water cooling is used and since the web is in contact with the chill rolls over a major part of their diameters it is possible to use them to meter the output tension of the web utilising PIV control. As the web is now dry, the chill roll stand can be a vertical assembly and a section can be enclosed to raise the moisture content of the web which has been lowered in the oven. The evaporation of moisture used in the chill rolls makes this easy to arrange.

Microwave drying

Microwaves are electromagnetic waves with wavelengths less than 200mm and frequencies between 1 000 and 12 000 megacycles per second. Like light waves they travel in straight lines, passing easily through some substances and being reflected or absorbed by others. They are being used increasingly in ink drying systems because they penetrate the whole depth of the ink layer to produce volume drying as distinct from the conduction drying which occurs with most other systems. With conventional radiation heating the outer surface of the ink layer receives the initial heat, which is then conducted through the ink film. The conventional systems are also more wasteful of their energy as they heat the air, the web and the machine in addition to the ink, often with undesirable side effects.

Although the theory of microwave drying has been well understood for some time, it is a comparatively recent development for web drying. A stringent relationship between the heater and the web path is essential as the web must pass within a few millimetres of the heating elements, and not vary from its path by more than a few tenths of a millimetre for maximum efficiency.

The source of microwaves is a set of magnetrons, each linked to a power pack which converts the AC current from the mains to DC. The power generated by the magnetrons is usually in the order of 5kW. A circulating water system and a heat exchanger are also necessary with this method of drying in order to cool the magnetrons.

If web control can be established accurately, microwave drying has a number of advantages. Pre-heating and cooling sections are not required in the system and moisture content of the substrate is hardly affected; dimensional changes are not such a problem, therefore, as with conventional dryers. Less space is required for this type of heating device and costs are competitive with other systems.

PRE-PRINTED WEBS

For high-speed, mass production publication work such as newspaper printing, it is often an advantage to print colour halftone work in advance in order to save valuable production time. This requires careful pre-planning of the publication and well organised co-ordination between the two runs, particularly where different processes are used for the colour and the text. This system goes against the principle that all printing should be done in one pass through the machine and a decision must first be made to employ one of two basic methods of producing pre-printed reels.

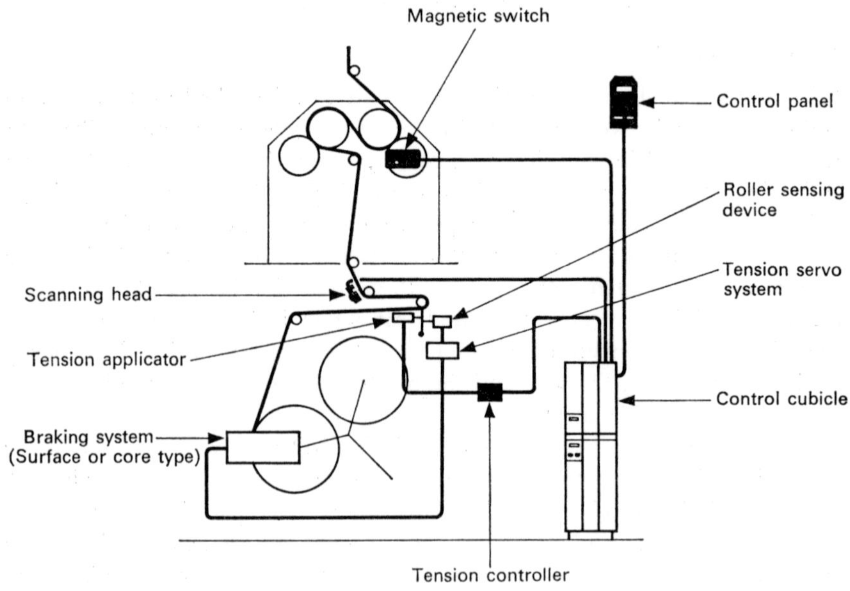

6.27 *Principles of insetting. A photo-electric scanning head generates impulses from register marks in the pre-printed designs. These impulses correspond to the position of the pre-print images as they enter the newspaper press. At the same time a magnetic switch, which is coupled to a printing cylinder, generates impulses which correspond to the position of the images being over-printed. Using the two series of impulses the insetter computes the condition and the rate of change of register. Should there be any deviation from the correct register the insetter makes automatic corrections by adjusting the degree of stretch in the pre-printed web.*

The easiest method is to print the colour as a continuous, repeating 'wallpaper' pattern so designed that it may be cut off at random during the final run without unduly spoiling the effect of the colour work. This system requires no special expertise on the part of the pre-print producer or the newspaper printer, other than the normal techniques required for producing their respective products. The main disadvantage is the difficulty of producing suitable patterned illustrations and the limitations of non-registering colour.

Where the colour printing is to be precisely registered to the news text, more stringent requirements are demanded of both printers. A first necessity is that the cylinder sizes of the pre-printing machine should correspond to those of the news machine. A tolerance of better than 0·1 per cent is usually expected, but it should be understood that even with this degree of accuracy the gradual creeping of a register variation over the course of a few thousand revolutions (a few minutes run) will be more than the conventional registering techniques can accommodate. For this reason it is essential to employ an automatic insetting register device for pre-prints which must register to the second run.

The principle of insetting requires the pre-printed web to stretch slightly as it is pulled into the news machine. The colour run, therefore, is usually printed a few hundredths of a millimetre shorter than the newspaper so that it can be stretched into register. The equipment cannot cope with a print-length which is oversize. Standardised procedures are essential for colour run and constant monitoring of the print length must be practised. Moisture content of the paper should be in the order of 7½ to 8½ per cent so that the web is sufficiently elastic to stretch without breaking. Variables such as running speed, pressure settings, ink viscosity and dryer temperature, should be stabilised as far as possible and recorded for future runs.

7. Ancillary equipment

There is a variety of available and necessary ancillary equipment for any printing system. This chapter considers such equipment in regard to quality control and measurement, static elimination, web-break detection, anti-set-off devices and operations such as bronzing, slitting, varnishing, etc.

QUALITY CONTROL

The quality of the end-product in printing may be measured or otherwise evaluated by reference to local, national or international standards. Some British Standards have been referred to in previous chapters, and a list of relevant standards will be found in the appendix.

An experienced print buyer will know precisely the standards he expects and when ordering work he will provide the printer with a specification. This is a very practical arrangement as it not only ensures that the buyer expresses his requirements exactly, but enables the printer to be sure that his work will be satisfactory provided he takes care to exercise appropriate control of his product.

The relevant standards will usually be related directly to the end use of the product; *eg* text matter must be legible, brochures faithful to colour reproduction, business forms must fit accounting machinery and packaging materials should have adequate strength. A standard may be a matter of agreement between buyer and printer, or both may agree to accept an established standard. National and international standards are particularly valuable where several firms are engaged in producing parts of a job, or work of a similar nature.

The most common quality control device is the 'copy'. This may be a model of the end product if it be something like a carton or folding box; a 'dummy' as made up for a publication such as a magazine, newspaper or book; or a proof of the printing plates where colour reproduction is critical as in the case of an advertisement or brochure. In each case the end product is not only specified, but a pattern is made for the printer to copy.

Phases of quality control

There are three phases of quality control; before, during and after printing.

Before printing materials should be checked for the standards required by the specification. For instance, papers may be checked for substance, colour, caliper, grain direction, surface strength, opacity, strength, smoothness, gloss, etc. Not all these checks will be essential for every job, but the basic idea of quality control before printing is to ensure that all materials are up to standard before the discovery of a shortcoming becomes expensive.

During printing frequent checks should be made for register, ink coverage, impression strength, colour values and the absence of the faults described in Chapter 5. This aspect is usually the direct responsibility of the machine printer, but it is advisable to take random samples for thorough inspection in the quality control section.

Post-printing checks include tests for ink drying, adhesion, bleed, material strength or shrinkage, print fading, blocking and every requirement detailed in the buyer's specification. This type of check need not necessarily be made too late for action to be taken. A proof may be made in the pre-printing section, simulating production conditions as far as possible, or a test run can be made before full production is commenced, so that possible troubles can be foreseen prior to running on.

Quality measurement

A printing press does not provide the means for compensating for random variations in ink or substrate. Ink film thickness and positional press control are largely dependent upon human judgement. Quality measurement is a means of improving rapid and intelligent control of those differences which will occur so that they may be kept within tolerable acceptance.

Positional measurement

Register tables. If platemaking includes positional accuracy between images and a method of image location, all flexible plates can be fitted to the machine with positional certainty using measuring devices. These can be available on the press, or separately arranged as a part of pre-makeready. Plate clamps in the first case need a system for synchronising with punch holes or register marks incorporated during platemaking. Register scales on the press allow for fine adjustment as required. The alternative system provides for plate clamps which slot into a fixed station on the press, but plates are pre-positioned in these clamps on a separate register table away from the machine. Micrometer adjustment and sighting devices are incorporated, and this system can also be linked to the registering methods used in platemaking

Multi-colour presses require a procedure for synchronising cylinders to ensure pre-accuracy of plate fitting. This may be obtained by bearer marks, photocells, or a controlled electronic drive system to each unit.

Web viewers. On high-speed web machines it is impossible to follow the printed image on the paper with the unaided eye. Where the work is delivered as cut sheets or folded copies, the print can be examined as it reaches the machine delivery, but there is always some delay in making corrections. The situation is even more difficult where the web is rewound, for the printer must either run the machine slowly enough to examine the work, or speed up and run 'blind'.

Web viewers are designed to overcome this problem. There are two basic types of viewer, but both apply the same optical principle. This is based on the phenomenon of persistence of vision, whereby the eye receives its information at the rate of several 'pictures' a second, and retains each image until it is replaced by the next. The idea is to take sequential images on the moving web and arrange for each one to be seen only in one position. If the successive images follow each other quickly enough, the eye will see them superimposed upon each other and therefore apparently stationary.

Stroboscopic viewers use a flashing light to achieve this effect. The moving web is illuminated by a brilliant light for a few microseconds when the image is in a certain position, and the next flash is made when the next printed image is in the same position.

The flashing light is produced by an instantaneous electric discharge between two electrodes in a gas-filled tube. A brief duration of the light is essential as the image must not move more than a few hundredths of a millimetre during the period of illumination. A quick calculation will show that at a web speed of 600 metres per minute the image will have moved 0·01 mm in one microsecond.

Rate of flashing must correspond to the print repeat length or a multiple of it. Again, at 600 metres per minute an image which repeats every 250 mm will pass the viewing line at a rate of 2 400 images per minute, or 40 per second: the flashing rate should equal this if the images are to appear stationary. If the flashing rate is slightly faster or slower then the image will appear to move slowly instead of standing still.

A triggering device must be connected to the tube to control the rate of discharge. It is possible to use a direct coupling to the machine where the stroboscope is used in a permanent position to examine a fixed print length. Where the print repeat varies from job to job the flashing rate may be controlled by a photo-electric cell scanning the web to locate a signal mark printed with each copy. Portable stroboscopes generally have a built-in variable flash timer, so that the operator can control the synchronisation manually.

Stroboscopes can, of course, be used to examine other moving parts with regular motions such as the printing units and folding mechanisms. They are most efficient where the flashing rate is above 50 per second, when the flicker caused by the intermittent light is less disturbing to the eye.

Mirror viewers. An alternative method of web viewing is to watch the print via a mirror system, designed to oscillate at web speed with a frequency related to the image repetition. The simplest of these systems employs one oscillating mirror, while more elaborate versions use a series of mirrors mounted on a drum which rotates to match the print repeat.

Although the web is viewed indirectly with this method it does have a number of advantages over stroboscopic inspection: 1, colour rendering is less likely to be affected than with the discharge tube of limited spectral range; 2, slower viewing speeds are possible as the mirrors move in harmony with the web instead of trying to 'freeze' movement completely; 3, a magnifying eyepiece can be used to examine the work critically; and 4, lighting can be made to order if required, *eg* transparent materials may be illuminated from behind to show ink density more clearly.

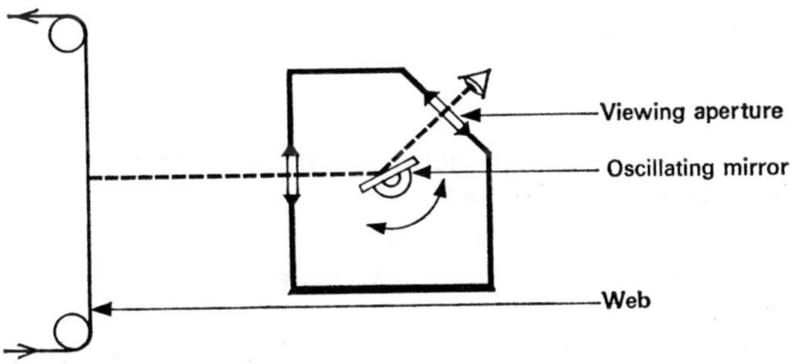

7.1 *Principle of oscillating mirror for viewing fast-moving webs. The oscillating mechanism can be adjusted to coincide with the print repeat on the web.*

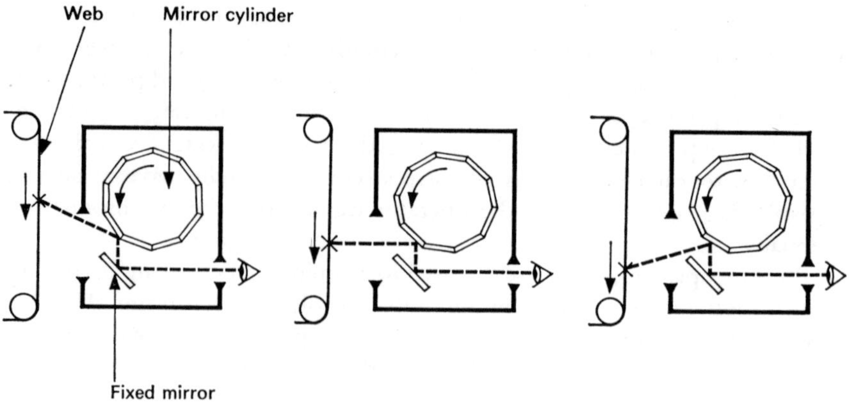

7.2 *Principle of revolving mirror for web viewing. The mirror cylinder is synchronised to web speed by selsyn control. The relative motions of web and mirror cylinder cause the reflected image in the fixed mirror to appear stationary due to the fact that the angle of the incident light will be equal to the angle of the reflected light. A slight interruption will occur as one mirror facet replaces another.*

Register controls

Equipments for the control of register are manufactured by a number of companies; a well known example is the Autotron. This equipment will control 1, length register of colours; 2, sideways register of colours; 3, back-up or perfecting register; 4, folder register; and 5, register of the web from the reelstand.

For the equipment to operate, it is necessary at the planning stage of a job to lay register marks in a suitable margin, preferably at the edge of the paper, where they will be ultimately trimmed off, or in the spine of the book. The register marks must be an exact distance apart and 10 mm long by 0·5 mm thick.

Scanning heads. A four-unit press printing yellow, magenta, cyan and black, will require three scanning heads; the first one mounted on the magenta unit and close to the outward nip will scan the yellow and magenta register marks. The other two are for scanning the magenta and cyan, and cyan and black printed marks.

The scanning head projects two spots of light on to the web in the path of the two register marks; as they pass the scanning head the reflected light is converted by two photocells into voltage pulses which are transmitted to a control desk. The signals are amplified and a special circuit calculates the error by multiplying the time interval between the pulses by the web speed. This error is shown at the control desk and at the control panel on calibrated meters.

Register motor. A correction signal is transmitted to a motor connected to a register roller which will alter the web length between printing units, or to differential gears in the drive to the printing cylinder and rotate it relative to the other printing cylinders. Many machine manufacturers use 'selsyn' motors which enable correction rates to be considerably increased by virtue of their precise action.

Magnetic switches. There is a magnetic switch for each scanning head. Each has a rotating disc carrying two small pole pieces. As the disc rotates the pole pieces pass across a coil and produce an impulse to trigger the scanning head which is live for 6 to 12 mm on either side of the register marks. Magnetic switches are geared to each unit or made into a composite unit and connected to the drive of the press to give a 1 to 1 ratio with the printing cylinder revolutions.

Control panel. This should be mounted where the pressman can see it. In the event of the job not being in register due to a register mark being out of position, it can be corrected to bring the printing image in register by moving the register-shift adjustment; this control is a duplication of one on the control desk. The control panel can be used to correct large errors and can then be switched to automatic.

Control desk. This is sometimes sited away from the machine; it is the nerve centre of the equipment and houses the transistorised printed circuits. Jack sockets are provided for testing the equipment and an oscilloscope will indicate whether or not the magnetic switches are correctly set. A short course of instruction enables the pressman to distinguish the different types of trace and their particular significances. The Autotron 100 has a predictor system in which the chassis contains a memory circuit which stores each error reading and compares it with previous readings. It will then predict the amount of error and make its correction. As this is a repeating process it prevents over-correction and brings the press into register quickly and accurately. Such equipment maintains a constant watch on press register and considerably reduces the printed waste rejected because of bad register.

A further development is obtained with the Autotron 130 mark recognition equipment. This dispenses with the magnetic switches. The scanning head must be accurately focused on to the printed register marks. The equipment contains a number of small photocells and an electronic logic system devised to recognise the register mark. A computer calculates the register error and the rate of change from signals transmitted by the scanning heads, and sends correction signals to the register correction motors. The rate of correction and degree of accuracy is the same as other models. Web speed compensation is provided by a tachogenerator coupled to the press drive. For split press work a second tachogenerator is necessary for the second delivery.

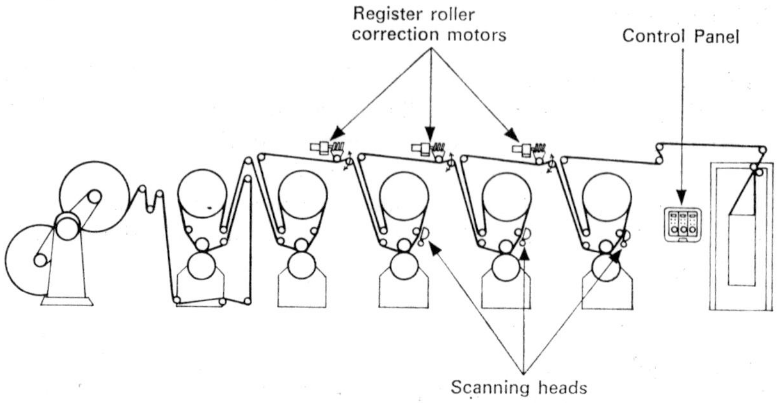

7.3 *Principle of automatic register control on web-fed machines (the Autotron system). The relationship between one printed ink film and the next may be controlled by photo-electric scanning as illustrated in this diagram.*

Press pre-alignment 135 is a development which enables a rotogravure press to be prepared, started up and running in register in a short space of time. The method of operation is that a register mark in the printing cylinder is used in conjunction with an optical system on the units to position the cylinders accurately. Partial braking is applied to the unit in order to take up backlash and tensional deflections for the final setting of the cylinder. Register rollers are set in a position to correspond with the cylinder size. For different size cylinders the register roller will be set in a different position and there are a number of requirements for ensuring that this procedure works with maximum advantage:

1 a barring motor for fine control of the line shaft angular position;

2 a cylinder coupling, which will allow the engagement of the cylinders at any angle;

3 the brakes must be capable of full or half power braking at the printing units;

4 web lengths between units should be multiples of the cylinder circumferences; and

5 the register mechanisms must be free of backlash and fitted with calibrated scales, and presses operating with various cylinder sizes should have linear register rollers.

INK AND COLOUR CONTROL

Equipment for this service covers three functions: 1, the need to establish specifications and tolerances before makeready; 2, to regulate control within these limits whilst running; and 3, to establish records of the run and assess quality measurement of the performance.

Proving comes into the first category since standards are set for subjective comparison. Conventional equipment may be used, or more frequently smaller proving machines with which the aim should be to produce a target within the capacity of an average production run rather than the achievement of a faultless proof. To obtain a more objective colour target, small proving instruments can be used with a known roller surface area to which ink can be applied in controlled increments so that progressive filmweights can be printed. For this purpose an ink pipette is used with an oscillatory roller system to shear the ink, and apparatus for preparing a print (BSI 4321 : 1969 refers). Knowledge of surface area of the rollers, coupled with calibration of the film splitting behaviour between rollers and substrate predetermines the precise filmweight of each print.

In connection with such preparation, the rheological characteristics of the ink may also be measured since any variation proposed to aid printing will affect the transfer characteristics of the ink. Falling bar or rotational viscometers are mostly used to establish viscosity and yield value of heavybodied inks whilst tack is separately assessed. The action of the printer who works his ink on a slab with a test of its resistance to his knife slicing through the heavy ink, is simulated by falling bar viscometry. Time of fall of a metal rod through a close fitting orifice lubricated by the ink enables plots to be made of time of fall against applied force. From these measurements a rheogram is drawn on a calibrated chart to enable the viscosity to be calculated under flow conditions, and the initial resistance to flow or yield value to be established. Rotational viscometers enable viscosities to be measured under higher shear rates, but are consequently less accurate with inks which back away from the rotating cylinder.

Tack is a less precise characteristic since it depends so much on the surface structures between which the ink film split is being made. The inkometer has been used to give a calibrated measurement of surface tack of inks but does contain an element of viscosity in its assessment. Consequently measurement of tack is usually made against a sample and is comparative. This applies to film split between rollers and onto the printing stock. A wide variety of machines and techniques are used to make this judgement. Automatic viscosity control is effected by a number of types of equipment for solvent-based inks as used for gravure, flexography and some coating applications. They range from purely mechanical devices to fully automatic electronic apparatus which meters the supply of ink, medium and solvent to the press.

Crosfield's Viscomex is a good example of automatic control. It consists of a viscosity motor or measuring element which is fixed to the top of the ink tank. The motor rotates a small cylinder in the ink, and with changes in viscosity there is a change in drag on the small cylinder which reacts on the current in the motor windings to form a basis for viscosity measurement.

A control cubicle houses as many control circuits as there are viscosity motors; each is sampled every 3 seconds and if viscosity error exists, solvent is added. A fail-safe device operates in the event of malfunction to prevent possible flooding, and at the same time the pressman is warned. Control cubicles are usually sited away from the machine and can feed more than one press. Control panels, however, must be easily seen from press deliveries. Each has a meter with graduated scale, control knob, switch for automatic or manual control, and an indicator lamp for that tank. The meter gives a direct reading of viscosity, and when the control panel is switched to manual, solvent can be added at a rapid rate necessary when preparing for running. When switched to automatic, the motor takes over and the lamp lights each time the solvent valve opens. In the event of a change in viscosity being required through a change in press speed or in substrate, the control knob effects this.

Solvent valves are either pneumatically or electrically operated and there is a handle for manual operation. The valve has a calibrated orifice which is adjustable to control the rate of flow to a particular unit. Similar valves are fitted for use with ink and medium and also incorporate the manual control handle. The solvent valve is controlled by an electrically operated air valve motivated by the control chassis and mounted as a separate unit beyond the flameproof area.

A level switch maintains the correct level of ink in its tank and the switch opens valves which supply ink and medium in proportions determined by the orifice settings. When the level has been raised by approximately 25 mm, ink and medium valves switch off so that the automatic viscosity equipment can then add the correct amount of solvent.

Automatic control systems provide replenishment in a matter of seconds working within fine limits of plus or minus one-half a centipoise. Other advantages are savings on ink through consistency, longer doctor blade life in gravure due to stable viscosity giving better blade edge lubrication, and a reduction in ink loss through material remaining in the tank at the end of the run.

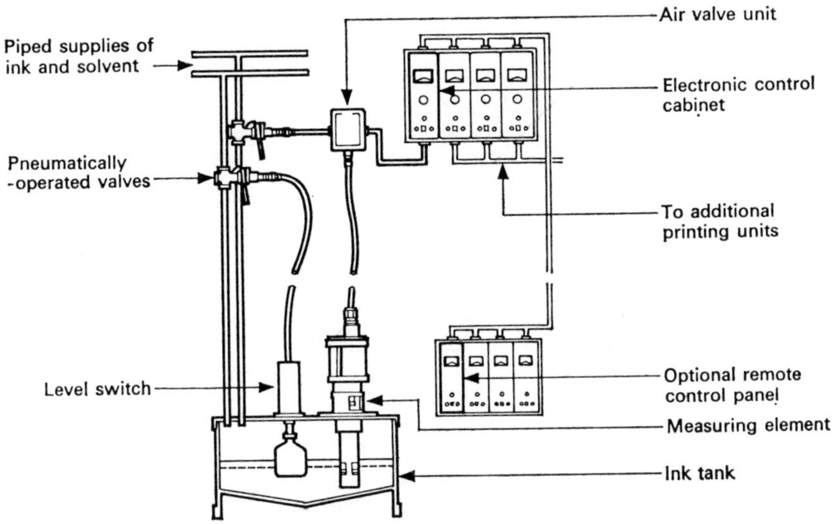

7.4 *Components of the Viscomex. The viscosity of a liquid ink is critical in determining its printing qualities and the proportion of solvent is probably the most significant factor which affects an inks viscosity.*

Instruments for colour measurement

Colour measurement and specification is carried out using a reflectometer responsive to light reflectance from the print surface, and there are three types of instruments in common use.

The densitometer measures reflectance in optical density units and is usually confined to a single density figure when used for colour control. Colour filters can be utilised to increase filmweight discrimination of the appropriate contrasting colour.

The colorimeter provides percentage reflectance through three or four specially computed filters to express red, blue and green content of the colour. A white standard is needed to establish maximum reflectance level and this is supplied with the instrument. A range of colour filters may be provided with such an instrument so that abridged spectrophotometric measurement can be made to identify colour records.

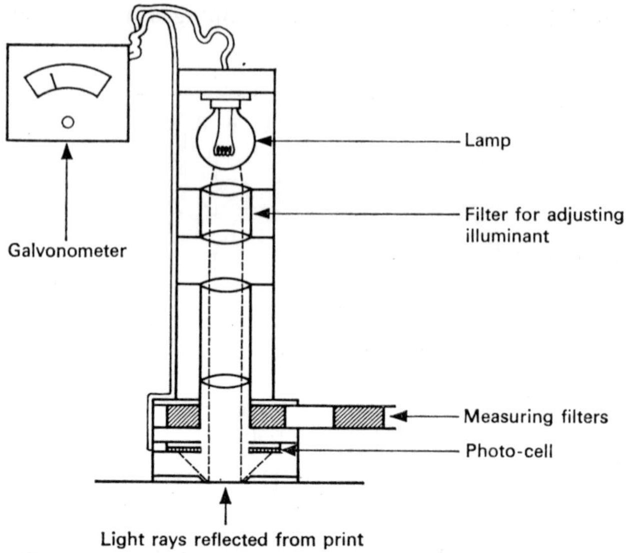

7.5 *Off-the-press colorimetry. Percentage reflectance is set to 100 on a white standard. Readings through suitable colour measuring filters give percentage assessment of sites on the printed area. These are plotted on the colour measurement chart, along with time of recording and passed to the printer for appropriate action. They can also be kept as a job record.*

The spectrophotometer is a more expensive instrument designed to provide accurate colour specification covering the range of the spectrum. Information is supplied as a continuous plot of reflectance against wavelength.

As an aid to the printer on the run, a densitometer or colorimeter with the appropriate colour filters sets target and tolerances accurately and notes trends towards increase or decrease in density before these become apparent to the eye. Statistical recording of a run is achieved by charting colour measurements. Quality measurement of completed work will show if dryback has caused any change from the measured wet impression. It should be noted here that instrument reflectance readings from a colorimeter or spectrophotometer are inverse to visual colour perception. Visual compensation in this technique should always be made, as reflectance has a logarithmic relationship to optical density, viz:

$$\text{Density D} = \log_{10} \frac{\text{reflectance of standard}}{\text{reflectance of sample}}$$

Metameric deviation is not measured by the above technique and must be avoided by standardising control of the ink mix during the run, and using a controlled light source for subjective judgements. A matching or viewing booth for this purpose may utilise balanced matching lamps to BS950 specification but should also incorporate a facility to provide alternative light source conditions such as tungsten and fluorescent variations. Many viewing booths provide only the British Standard lighting, and whilst this is satisfactory for objective comparison or assessments, the human eye needs the authority of alternative viewing illuminants to prevent incomplete colour matching.

Electrostatic assist. The Gravure Research Association in America established that electrostatic assist improved the print quality when applied at the nip of a gravure printing press. The Heliostat is an instrument developed for this purpose.

For many years gravure printers have been faced with a problem known as 'skip' or 'speckle'. This is the failure of cells to print out properly, particularly in the highlight areas. In designing electrostatic equipment to prevent this problem, safety, efficiency and economy must satisfy both printers and the authorities.

The Heliostat is a special sealed roller containing a small generator, the frame of which is rotated by the roller. In earlier models the shaft was prevented from turning by an eccentric weight but in later models a clutch principle is used. Also inside the roller are a transformer and appropriate electronic circuits. The voltage output is connected to the conductive rubber surface of the roller, and can be set from 500 to 2 000 volts by a knob on a pneumatic control panel and may be adjusted while the press is running.

The voltage can be checked by a push button which lowers a voltage pickup roller on to the Heliostat roller and if necessary the roller can be switched off on the panel. On the end of the roller journal is a polarity switch for use in dealing with inks of differing polarity; (unlike poles attract, like poles repel). If the wrong polarity is used there will be a refusal of ink to transfer.

The voltage used will depend on the thickness of the substrate. In general the thicker the material the higher the voltage requirement. The Heliostat runs in contact with the machine impression roller, which must have a covering of conductive rubber and the current is only generated when the press is running. Back up and bowl rollers that come into contact with the impression rollers need to be insulated in order to prevent the charge from earthing.

Because of the improved ink transfer some papers that are otherwise unsuitable for gravure can produce good results. Printing pressures for the electrostatic assembly are about one-third of those in normal use which means less wear and tear on rollers, machine bearings and gears.

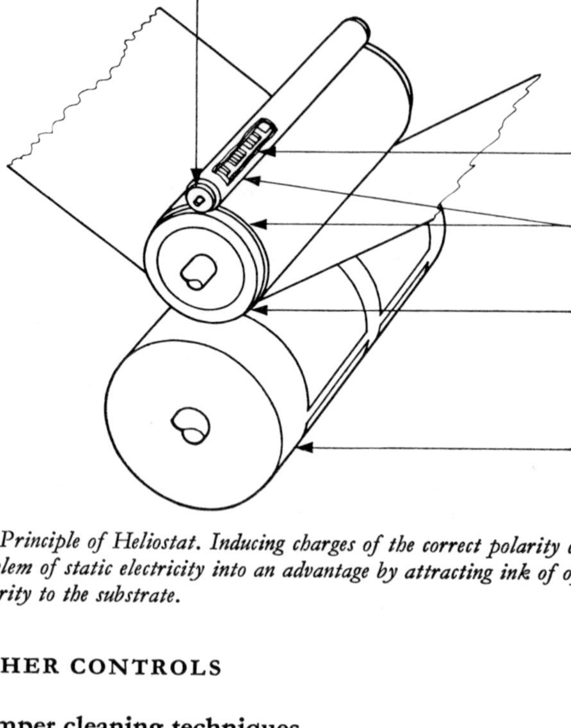

7.6 *Principle of Heliostat. Inducing charges of the correct polarity can turn the problem of static electricity into an advantage by attracting ink of opposite polarity to the substrate.*

OTHER CONTROLS

Damper cleaning techniques

Unlike the inking rollers, dampers are cleaned off the machine and where uniformity of damper sizes is maintained a replacement service can be operated if there are sufficient machines to make this worthwhile. Cleaning machines afford a mechanised method of scrubbing molleton-covered dampers, and use a rotary action to apply desensitiser, solvent, and warm water in turn. Low reservoir damper-covering materials such as stockinette do not require such attention and are cleaned on the machine or replaced. All solvent used for cleaning must be removed before the press run as solvent remaining in damping equipment will break down the pigment envelope in the ink film causing dewetting or blinding of the printing plate image.

Control of solution additives

Additives to the fountain solution are controlled by the use of pH monitors. Three additives are concerned:

1 the desensitiser which may be gum arabic, or cellulose gum (CMC), with a preservative usually included;

2 an acid such as phosphoric acid to improve adherence of the desensitiser to the plate by converting it to its free acid form and to degrease the non-image areas when scum is formed; and

3 a buffering material to slow up the corrosive activity of the acid and give a reserve of constant pH.

For the purposes of manual control, gum concentrate and etch concentrate are made up separately, the latter containing acid and buffer. Gum additions to a fountain should not exceed a specific limit or desensitisation will spread to image areas. Etch concentrate may be varied in acidity according to plant conditions and printing surface in use. Measurement of acidity or pH may be by using indicators which change colour at certain pH levels, or by electrical methods which depend upon the change of voltage at small electrodes.

Automatic monitoring systems measure the fountain solution in a circulator system using turbulence and filtering to assist in keeping a constant pH level. Control is over the balance between positive and negative ions, and careful technique is necessary to ensure that buffering does not vary, nor optimum desensitising levels become exceeded. In an automatic system, therefore, the predetermined percentages of acid/buffer/gum should be controlled. Circulation and filtering help to keep the solution biologically clean and pipes free from corrosion. Refrigeration may also be included to control fountain temperature and prevent water evaporation increasing fountain acidity.

Alcohol additives can be metered in a similar way to pH concentrates, either for use with conventional or unclad damper systems, or when using a continuous feed system such as the Dahlgren. Monitoring is less complicated in this instance since normally only the two wetting agents are intermixed: variation being a consequence of the evaporation of alcohol.

Quantitative measurement of press damping can be made by monitoring surface tack conditions on the ink train, or by utilising an infra-red photometer to measure the absorption spectrum of moisture on the plate surface. The second method is in the development stage, but the first has been in use for a number of years. It is useful particularly when printing non-absorbent materials such as tinplate which require low damping levels and fine control to aid trapping problems when printing multi-colour.

Static elimination

Under certain conditions non-conductive materials, such as papers or plastic films, can capture an electrical surface charge which will be unable to leak away through the material or the atmosphere, which also is normally non-conductive. The charge may be generated by friction or by rapid separation of two surfaces, resulting in an imbalance of electrons to give a positive or negative charge. Conditions for the generation of 'static' exist in most feeder sheet separators or reel unwinds. As the material passes through the machine further friction or separation may increase the problem and the charges may be transmitted to other parts of the machine by induction.

Due to the fact that unlike electrical charges attract each other and like charges repel, the passage of a charged material will give rise to a number of difficulties. Sheets may stick together or repel each other, and the material may also be attracted to or rejected by parts of the machine. In severe cases sheet or web control will be impossible and the material will be unusable unless the charges can be neutralised or dissipated.

Any approach to this problem must be aimed at allowing the captured charge to flow through a conducting medium. For instance, although dry paper is non-conductive, water is: and if the paper contains sufficient moisture it will conduct the charge. Generally speaking, papers with moisture contents of more than 5 per cent by weight will allow charges to leak away. Machine room and paper store conditions will influence the moisture content of the stock. RH should be at least 60 per cent in these rooms and the paper should be 'conditioned' to the room situation.

Ideal atmospheric conditions and material conditioning are not always possible, however, and even when they are practised may not be a universal cure for static. For instance, plastic films have no moisture content to help charges to leak away. Under these circumstances a different approach may be used whereby the charge is released by providing an artificial conducting path for the trapped electrons.

A simple method is to coat the charged surfaces with a conducting medium. Atomising sprays can be used which contain surface-active agents. Spraying with these aerosols will result in the material being coated with a very thin conducting layer. Before using this method it should be established that the spray will not affect the printed result or any of the end-product's desired properties.

Ionisers. Maximum efficiency is obtained with the use of electrical eliminators which ionise the air in the vicinity of the material, allowing the charge to leak away.

These 'neutralisers' comprise a transformer giving an alternating secondary potential of several thousand volts at a number of points, while reducing the current to a fraction of an ampere. The device is located as close as possible to the trouble area. The electrical fields thus created alternatively provide a positive field to neutralise negative charges and then a negative field to neutralise the remaining positive charges. The effective result is a layer of conducting air around the material which allows the charges to leak away.

A similar ionising effect can be obtained by using a radioactive material instead of the electrical device. The charged particles emitted by radiation create a conductive layer of air to neutralise the charges on the surface of the material.

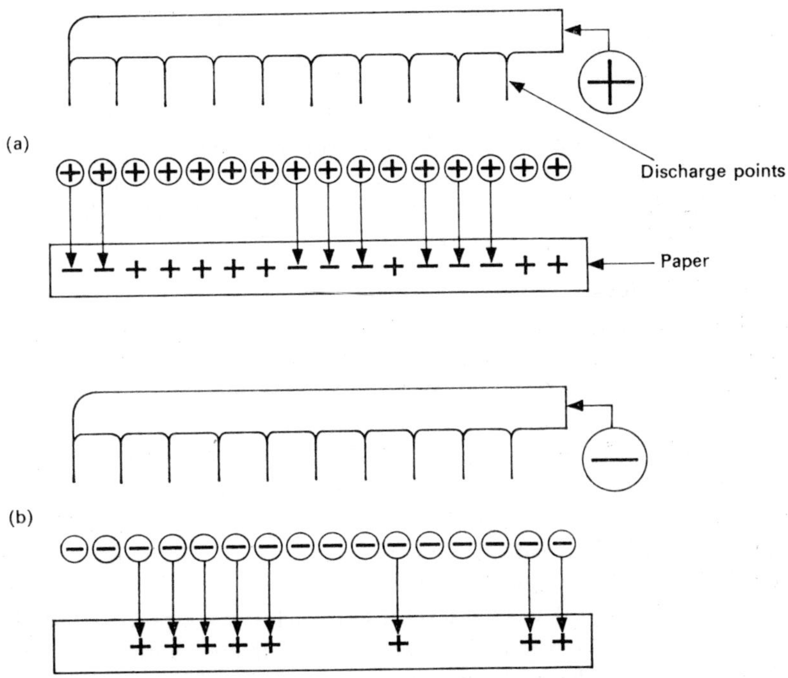

7.7 *Principle of static elimination by electrical discharge. (a) A high alternating potential fed to discharge points first creates a positive field which neutralises the negative charges in the substrate. (b) A negative field then neutralises the positive charges.*

Web-break detectors

On high-speed machines, especially with several webs, it is essential to stop the machine immediately in the event of a web break. Failure to stop quickly will mean that the web must be re-threaded completely through the machine, and uncontrolled lengths of paper are likely to be diverted around printing cylinders or choke up folders and other mechanisms. At top speed a printing unit could wrap a hundred turns of paper around one cylinder in less than five seconds, and long before this situation was reached the plates and cylinder dressings would be ruined, with probable permanent damage to the machine itself.

Detectors fall into two basic categories; mechanical and photo-electric. The mechanical systems are quite simple devices, comprising pivoted fingers which rest on the web, and in falling from their positions as the web breaks, energise a relay which stops the machine and operates the machine brakes.

Photo-electric systems do not require web contact. A light beam is projected on to the web and when a break occurs a photocell behind the web receives the light and energises a relay.

Whichever method is used, it should be incorporated into an overall system which not only stops the machine in the event of a break, but controls other functions such as preventing the reels from overrunning, operating web divertors which prevent loose paper from entering critical parts of the machine, shutting off the oven dryers, and throwing apart vital cylinder nips which may be damaged by paper jams.

Anti-set-off devices

Set-off is the unwanted transfer of ink from the print to another surface. The earliest approach to counter set-off was to interleave heavily-inked sheets of work with a rough-surfaced paper to keep them apart and prevent ink transferring.

Quick-setting inks and accelerated drying systems have largely reduced the need for anti-set-off precautions, but there are still occasions where a full film of ink could cause set-off, particularly where the work is delivered in deep piles, where multicolour work is done in one pass through the machine and if the printing stock has low absorbency.

Modern anti-set-off devices for sheet-fed machines are powder spraying systems, designed to cover each sheet as it is delivered with the thinnest possible layer of powder, to keep it from contact with the next sheet.

The powders used are generally a form of edible starch or calcium carbonate, with anti-clogging additives. The average particle sizes vary from 14 micrometres for the thinnest ink films such as are used in lithography, to

100 or more micrometres for the heavy ink films carried in some types of packaging work.

The simplest systems are little more than dusting devices in which low-pressure air is passed through a powder container to a nozzle which puffs powder at the sheet. Better systems will meter a specific quantity of powder according to the job requirements, and a purpose-made compressor will give better control of air pressure – between 100 and 500 newtons per square metre.

The minimum amount of powder to prevent set-off should be used. Too much powder, overall or locally, may affect the finish of the ink, cause filling-in, prevent other colours from overprinting, or hinder post-printing operations such as varnishing or laminating.

The simpler devices spray the sheet from one or two nozzles. It is not always possible to guarantee an even distribution of powder from a central position or from random positioning of such spraying heads, particularly when it is realised that the amount of spray used for most work will be hardly visible when the device is working. A better system is to use a spray bar, the

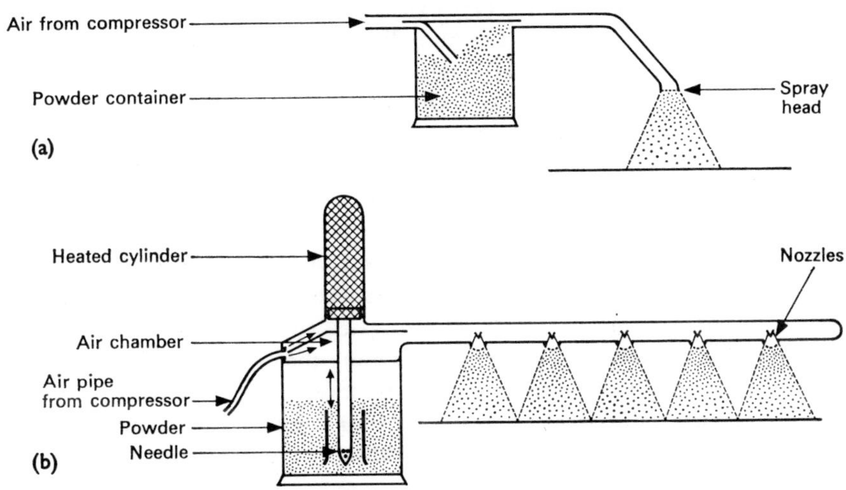

7.8 *Anti-setoff sprays. (a) A low pressure air supply is divided into two channels, one to agitate the powder in the container and the other to carry the powder to the spray head. (b) This system provides more precise control and better coverage. The needle has a number of small openings to contain a measured quantity of powder. The charged needle is lifted to the air chamber by air pressure and the powder discharged to the nozzles. The needle is kept warm to avoid clogging of the particles due to moisture.*

full width of the sheet, containing a row of powder outlets to ensure overall coverage without missing corners or double-spraying.

A more sophisticated method utilises the electrostatic transfer of spray powder to the printed sheets to give a finer but more even distribution and eliminate the wide dispersal of spray powder that occurs with air blasting systems. The corona of ionised air is also useful in eliminating static electricity from the printed stock to allow perfect jogging of sheets and an even stack in the delivery. Absorbent edible starch powders are used, but a resin coated hydrophobic powder is available if required.

Web machines can be fitted with spray bars if desired, but it is generally more efficient and economical to ensure thorough drying of the ink by means of an accelerated drying system.

Paper conditioning

The range of laboratory paper tests is considerable. The printer will normally confine his own tests to assessment of proofs to confirm ink/paper drying relationships, rub testing etc. He is concerned, however, in the physical condition of the stock and the need to maintain correct moisture equilibrium. Sword-type instruments enable paper stacks to be tested to confirm the relationship of their temperature and humidity with ambient conditions. Stack temperature should be allowed to rise to room condition by early delivery to the appropriate department. The sword hygrometer will then indicate the relative humidity in the stack which ought to be 5·8 per cent above room conditions. The stack instrument can be inserted into a wrapped load and stacks should not be unwrapped before a moisture balance check is made.

Paper conditioning equipment enables unsatisfactory loads to be brought to room conditions quickly. Paper is hung in small quantities in a fixed or travelling cabinet containing blowers and humidifiers, controlled to bring the stock into room equilibrium. Sufficient space should be allowed for free circulation of air, and stacks inverted and hung the other way after completing half their conditioning time. Sheet substance, size and moisture differentials decide the conditioning time. Effective air circulation is the most important essential, and requires that lifts should not be overloaded.

ANCILLARY OPERATIONS

Varnishing

Varnish is a clear liquid which is applied to a substrate as a final coating so that, upon drying, it enhances the appearance of the finished product and imparts protective qualities. The varnish may be applied by specialised equip-

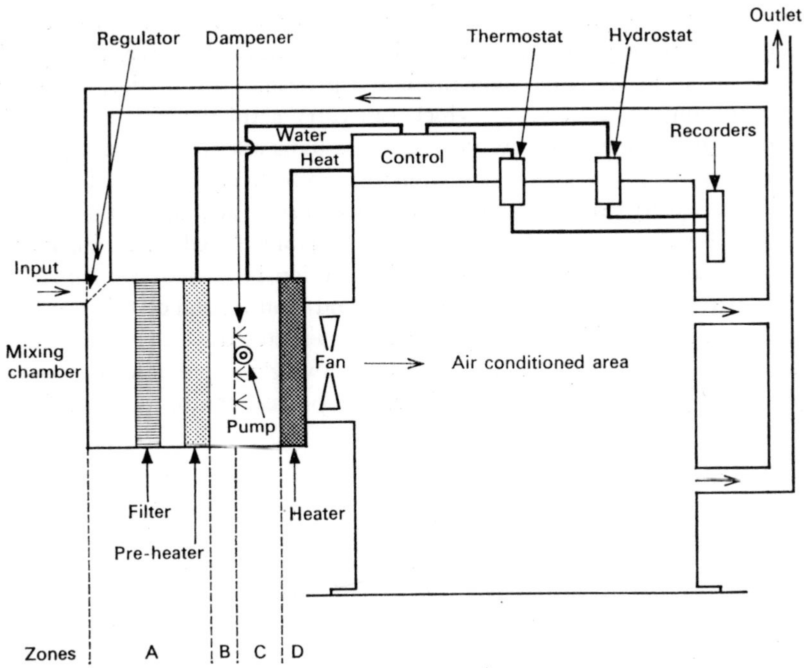

7.9 *Full air conditioning plant. Zone A: re-circulated and fresh air, mixed and filtered, are fed in at controlled temperature. Zone B: air is cooled to its dewpoint and water circulated. Zone C: further cooling along a line of saturation condenses out superfluous water and fixes new dewpoint. Zone D: air is warmed to the control temperature and circulated.*

ment in the print finishing department or by a printing machine as an additional operation. The coating may be over the whole of the sheet or, if some areas are to be left unvarnished, in a pattern similar to a printed image. Any of the main printing processes can be adapted for this process.

Surfaces to be varnished should be smooth in the first instance if a good gloss is required. Varnishes will vary in their formulations according to the process used and the degree of gloss and other properties required in the end product. Their working properties should be similar to those of the related printing process and they should be checked for compatibility with the printing inks and the substrate. Particular problems are bleed, which may take some time to show; and discolouration upon extended exposure to the atmosphere.

As coverage is usually overall or a very simple pattern, a hand-cut plate may be used for application by letterpress. For offset production, the normal lithographic principle can sometimes be used, but as this gives a very thin film resulting in low gloss and minimum protective properties, it may be better to employ relief plates which can carry a thicker film without the complication of water. Where very accurate films are necessary the gravure process may be preferred as the etch depth can be finely controlled and with accurate viscosity control plus correct choice of solvents, rapid drying can be achieved. For these reasons gravure is often used for web-fed varnishing inline.

Spirit varnishes consisting of a resin solution in alcohol are applied by roller coating methods to give high gloss and scuff resistance allied to other properties. Application is simply effected by a metered roller rotating in the coating material. Minimum control over filmweight is possible, and reverse rolling will produce an enhanced gloss. The process can be automated with

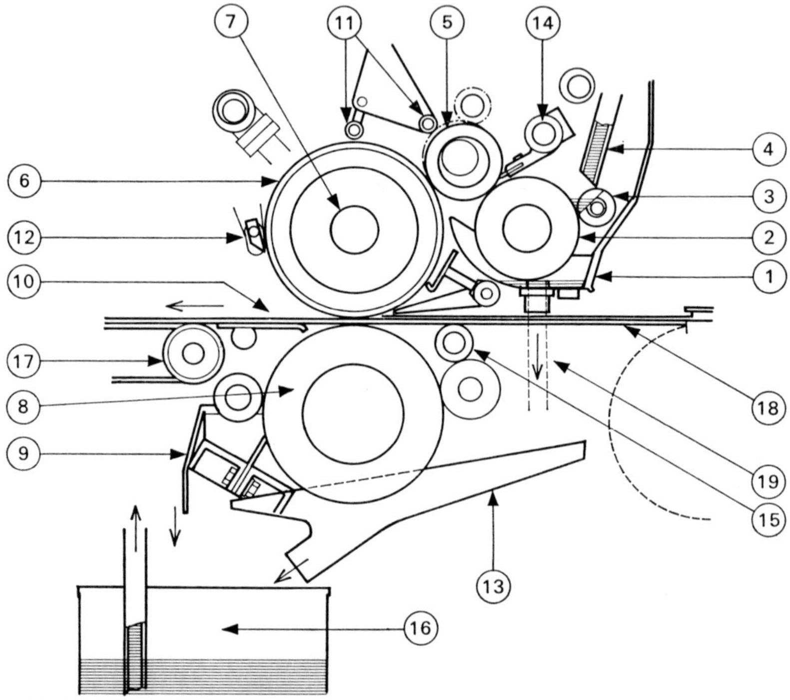

7.10 *Mailander stencil coater for metal decoration. 1, lacquer duct. 2, lacquer duct roller. 3, lacquer throttle shaft. 4, lacquer supply. 5, transfer roller. 6, application cylinder. 7, replaceable shaft. 8, back pressure roller. 9, scraper bar. 10, sheet delivery (to oven). 11, safety bars. 12, stripper device. 13, lacquer recirculation pan. 14, scraper for stencillings. 15, friction rolls. 16, lacquer container. 17, conveyor belt. 18, registered sheet. 19, return pipe to supply.*

similar feeder and stacking procedures to those in use on printing machinery and in-line ovens may accelerate drying by forced air circulation, and/or electric radiant heat.

Metal decorating

Roller coating machines for metal decoration require finer control over film-weight than is necessary for paper varnishing. Apart from the greater cost of materials involved, accurate control is necessary to achieve constant performance in fabrication, adhesion and protective qualities of the dry film. With simple roller coating machines, micrometer adjustment between two steel rollers meters the varnish, which is applied directly by a rubber, gelatine, or polyurethane surfaced roller. By the same technique, protective lacquers and opaque base coatings can also be applied to give protection and decoration to metal articles and containers.

Utilising more sophisticated machinery, stencil application can be made to register and impose with printed work. Blanket mounting cylinders are fitted for this purpose, and stencils may be cut on the machine, or on a separate cylinder or table jig incorporating bias measuring devices to compensate for tangential error.

Metal coating materials are 1, base sizes to give good adhesion between metal and the decorative coating; 2, high temperature gold stoving lacquers mainly for internal can protection and external corrosion resistance purposes; 3, pigmented roller coatings as base for print decoration or on their own; and 4, finishing varnishes. These will be used as required for high pressure containers including cans for most processed foods, and general line work printing low-pressure containers or decorated ware. Coating materials are oleo-resinous or synthetic and range from high temperature thermosetting varieties to low temperature thermoplastic coatings with charring tendencies if overstoved.

Metal decorating ovens

Decoration to metal surfaces has to be stove dried upon the surface. Standard conditions are essential to avoid colour variation occurring at varying temperatures, and to ensure coated materials have the right film characteristics. Ovens are normally gas fired and fulfil four basic functions: 1, to raise the metal temperature to specified requirements in the shortest practical time; 2, to maintain it at that temperature for the specified time; 3, to exhaust volatiles given off, and 4, to re-cool the metal before stacking. Since the oven is linked to the press drive, machine speed directly influences the stoving time and must be controlled within drying tolerances. Matching of oven dimension and type of work, with the ink and coating performance, enables standard production speeds to be maintained.

Efficiency in heat transfer within the oven is necessary since the volume of metal to be raised in temperature includes the racking system on chains in the travelling oven. Air circulation velocities are therefore high. Where economical, external firing from a furnace box provides hot air circulation through the oven. Where heat loss is considerable, as on very large assemblies for drum printing, internal convection firing is used. Electrical and oil fired systems are also in use to a lesser extent but have disadvantages compared to town or natural gas. An aspect of all ovens is their effect on plant conditions. A metal decorating oven exhausts to the outer air and input should similarly be drawn from outside if draught conditions in the factory are to be avoided.

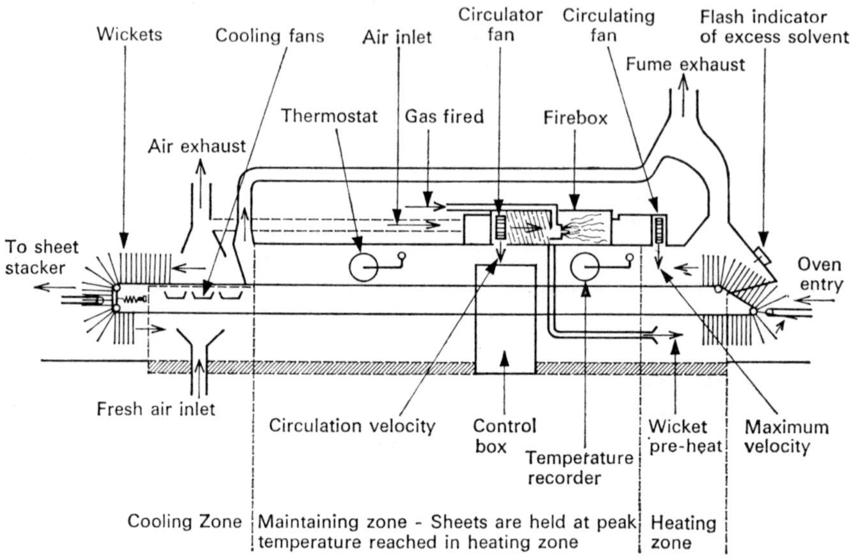

7.11 *Direct externally-gas-fired metal decorating oven. When printing on metal it is necessary to ensure adhesion of the inks and coating media by high-temperature stoving in an oven such as the one shown here.*

Bronzing

This is a technique whereby metallic powder is applied as an in-line operation to printed size or glue and then burnished. The size print must be sufficiently tacky to enable the bronze to adhere and may range in tone according to the colour of the bronze applied and the effect desired. The machinery for applying the metallic powders must be adequately enclosed since the particles are

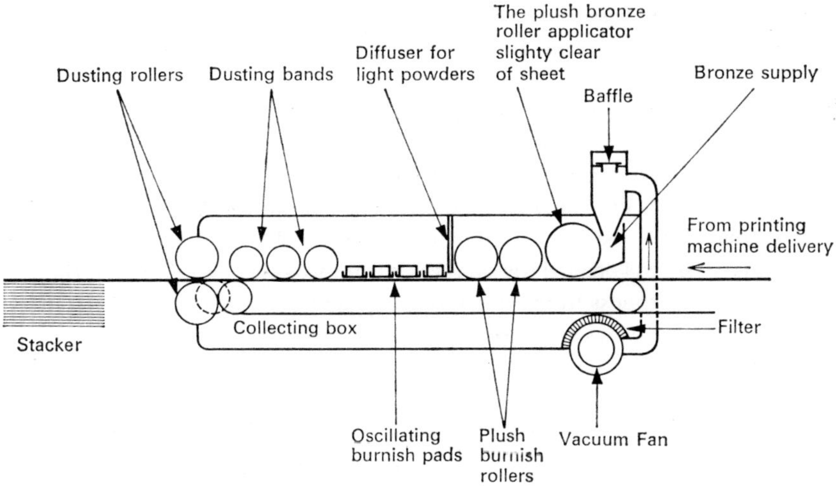

7.12 *Bronzing machine. Metallic powders are applied to a printed or sized base pattern on the substrate in order to produce a metallised printing surface.*

a health hazard. The mechanical operation involves dusting the metallic powder onto the wet print by brush and band techniques which alternately dust off and burnish. Burnishing may be made more effective as a postprinting operation by calendering methods which flatten the metallic surface to provide a smooth finish.

It is essential that previous printing on the paper stock be completely dry, or metallic powder will adhere in the wrong places. The printed tacky size must overprint well if necessary. Ideally it is better if the adhesive is not underprinted since it is difficult to cater for a differing absorption factor over the sheet. Bronze problems may occur through the presence of static charges which must be dissipated. This is most effectively done by spraying antistatic compounds onto the dusting bands at periodic intervals. Sulphur is an enemy of most bronze powders and responsible for tarnishing. Papers showing above three parts per million of reducible sulphur should not be used for bronzing work.

Slitting

There are two main reasons for slitting a web. One is to divide the web into narrower strips and the other is to trim the edge to a required size and make a neat finish to the job.

Slitting can be done by sharpened steel discs, cutting through the paper on to a steel cylinder. A sharp, clean cut is produced and the discs are spring-loaded to avoid excessive wear. Even so the discs must have regular maintenance to ensure sharpness, and excessive pressures will lead to scoring of the counter-cylinder.

Less maintenance is required if two flat-sided discs are used for each slitting position. The two discs are set, one above and one below the web and run against each other in a rotary scissor action. The double-disc system takes a little longer to set up but is trouble free when running. Another alternative is to arrange for the slitter disc to run in a slot in the counter-cylinder. Very little maintenance is needed for this system and wear is negligible, but there is some loss of sharpness in the cut and the arrangement is only really suitable where the slit is always required in the same place.

Perforating

Length perforating is done with sharpened discs similar to slitter wheels but with small sections removed to give the perforating action.

Cross perforations are made by a toothed blade held in one cylinder cutting against a hardened steel surface in a second cylinder, or alternatively, into a slot. The 'hard-on-hard' system is more versatile and provides a sharper perforation, but as with hard-on-hard slitters, wear takes place more rapidly. Slot-type perforations limit the positions of the perforations and tend to emboss the paper on the reverse side, but the blades and counter cylinders last much longer.

Hole punching

Postage stamp perforations are really sets of small holes punched in the paper. These are made by sets of steel pins penetrating through the paper into a die plate. The dies must be manufactured of hardened steel to fine tolerances. Adequate provision must also be made for ejection of the waste material, which is punched into the die-holes, to avoid clogging and damaging the pins. Pins may be slightly less hard compared to the dies, as they may be gently spread by tapping the ends when wear occurs. This is a fairly delicate operation and can only be done a limited number of times.

If the holes must conform to a regular pattern of size and spacing, as with stamps or the marginal holes used for feeding continuous stationery, the punching unit can be designed as a special unit. Where variations in hole size and spacing must be provided for, the units must be designed so that punches and dies can be easily changed and adjusted for position to suit the changing needs of the work.

Rotary units for hole punching must meet the same requirements as flat ones, but their action is based on a different principle. Flat units have a true punching action and the force needed to cut the holes is proportional to the total contact area under shear. In other words, the bigger the holes and the more of them, the greater the power needed. With a rotary unit the holes are made by the edge of the punch entering the die first and the rest of the punch following through, so that a cutting action is made rather than a punch. Consequently, little more power is needed to cut larger holes or greater numbers.

The cutting action of rotary punches is more complicated with larger holes. Punches of a few millimetres in diameter may be steel pins with flat punching faces and the dies can be small, flat plates with holes of corresponding size, locked into the die cylinder to make a many-faceted drum. This system can be used successfully where the hole diameter is less than about 1 per cent of the cylinder diameter, but larger punches must be made with their cutting faces curved to the same arc as the cylinder which carries them, and of course the same principle applies to the dies. It can be seen that the design and operation of units for punching larger holes is much more complicated and costly than for smaller ones.

Numbering

Consecutive numbering is required on several types of printed product and specialised equipment is needed to do this work. A numbering box contains a set of polygonal wheels, each wheel having ten digits cut in relief on the exterior facets. A ratchet and pawl device operates the wheel-turning mechanism so that at each impression the first wheel rotates to the next digit. At every tenth move the second wheel is engaged and turned to its next figure, while ten moves on the second wheel operates the third, etc.

For printing purposes the numbering box may be considered as a normal relief surface except that it changes its image every impression. Its requirements for inking, impression and registration are the same, and for flatbed work the numbering box may even be included as part of the forme.

The pawl movement for flatbed boxes is usually operated by a plunger which is depressed by the cylinder or platen at every impression. On rotary machines the pawl is connected to a cam roller which runs in an elliptical cam path to actuate the wheels.

The printing face of a rotary numbering box should be suitably curved to the arc which it describes. This is rarely done where the numbers run parallel to the axis of the cylinder with the result that an even impression is difficult to obtain with the decagonal wheels. This is not always so much of a disadvantage as it may at first seem, and in practice the same boxes are often used with a wide range of different cylinder circumferences.

Where the numbers on a rotary unit run around the cylinder, however, it is essential that the face of the box conforms to the correct arc. For this reason the boxes are constructed with barrel-shaped sets of numbering wheels. The work entailed in designing and making the barrel boxes obviously makes them more expensive and they have the added limitation that they can only be used in conjunction with cylinders of one diameter.

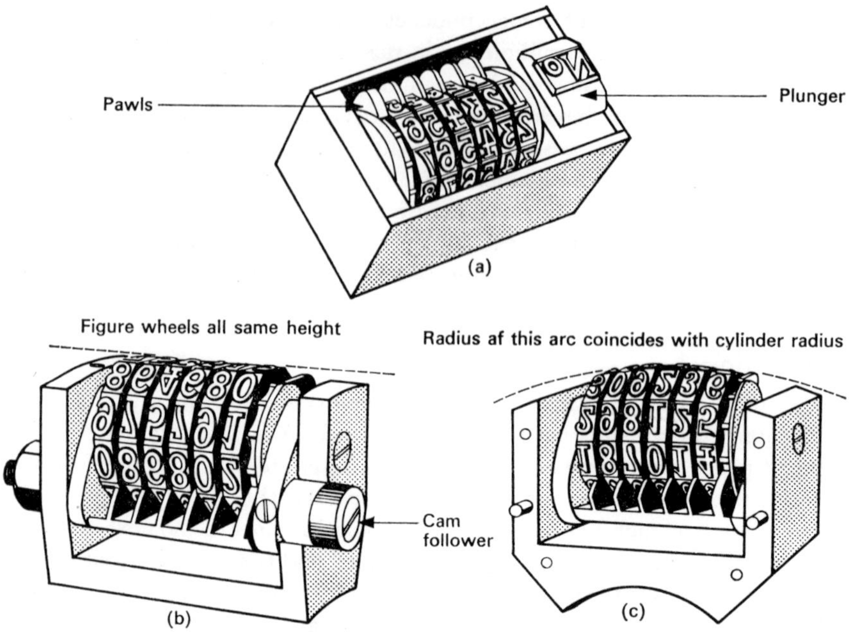

7.13 *Numbering. (a) 6-wheel numbering box for flatbed printing. The plunger is operated when the prefix (N°) is depressed by the cylinder or platen. Each impression turns the first wheel to the next digit and an interlocking pawl mechanism turns the second wheel after ten moves, etc. (b) Numbering box for a rotary machine. This box is operated by a roller following a cam in a specially-designed cylinder. The box is only suitable for numbering across the web direction. (c) Box for numbering in the web direction. Each wheel must be curved to the correct arc for the cylinder and a set of wheels will appear barrel shaped.*

Magnetic ink numbering

There are a number of business systems which are able to recognise symbols on printed documents. Some systems recognise the characters optically, as the human eye does, while others can interpret small magnetic signals from specially shaped figures printed with ink containing a ferrous material.

The names given to these systems are respectively, Optical Character Recognition (OCR) and Magnetic Ink Character Recognition (MICR). Magnetic systems are generally concerned primarily with numbers for use with rapid sorting systems. Most British cheques now use numbers printed in MICR characters for speedy processing at the clearing houses.

The characters for MICR are specially designed to give easily recognisable signals to the reading and sorting device. Some have a superficial resemblance to conventional numbers, while others are exclusively designed to serve the sorting purpose. Special numbering boxes are made for MICR work and printing from them needs to be meticulously carried out as the slightest variation in print quality can alter a character's signal value.

Impression needs to be perfectly even and firm so that the correct character shape is printed without missed areas or squeezed edges. This requirement may be particularly difficult to meet on rotary numbering units if the printing faces of the digits are not curved to the arc of the cylinder.

Inking must be equally good, ensuring that a sufficient film of ink is printed to give adequate signal strength, but not too much ink to cause squash at the edges. Roller settings must be very precise to avoid wipe.

Positioning of the numbers must be accurate to within 1 mm as misplacement of the characters could result in misreading at the sorting end.

Quality control equipment for magnetic ink printing may include transparent, squared grids for position checking; enlarging, optical comparators for establishing the fidelity of the symbol shapes; ink film thickness monitors; signal strength readers; and signal shape comparators which translate the signals into wave forms on an oscilloscope. All of these will be in addition to the normal printing quality control equipment.

The inks used for MICR printing have been dealt with in chapter 3.

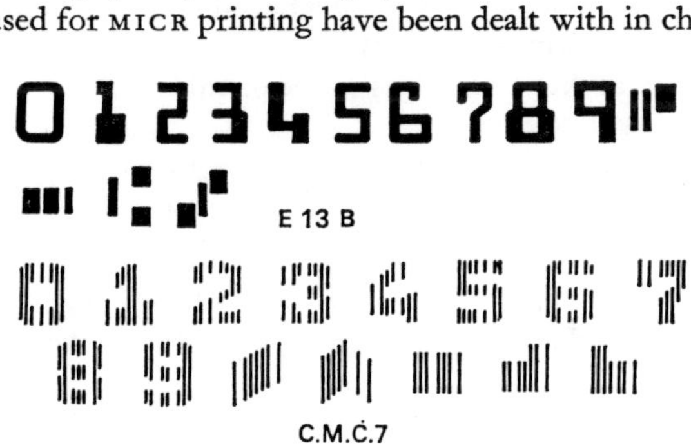

7.14 *Magnetic ink printing. Special characters are used for numbering in conjunction with high-speed sorting systems. Two of the most common styles are shown here.*

Cutting and creasing

Where a sheet is to be scored, creased, cut or perforated, a letterpress machine may be used to perform the operation by including a rule, or rules, of the appropriate sort in the forme. A separate operation is usually necessary as the rules will cut the inking rollers if used in the printing forme, although there are some patent systems which will automatically depress the rules below forme height as the inkers pass over them. A selection of hardened steel rules is available for each operation and the techniques employed will differ in each instance according to the effect required.

Scoring is done by a sharp-edged rule in the forme which is set to the stock by careful adjustment of pressure so that it will compress the fibres along the line of the rule and cut partially through the material. This will allow the sheet to be folded readily in the direction away from the side which the blade contacts. The fold may not be of good quality, but the operation is fairly quick and simple to arrange.

Creasing. A better fold is obtained by using a slightly rounded edge to the rule and using it to force the stock into a channel formed by two boards on the platen or cylinder. This method provides a correctly shaped bolt or hinge so that the sheet folds easily without gaping and if properly performed the hinge should actually strengthen the fold, whereas scoring will weaken it. The caliper of the sheet is significant in determining the width of the channel into which it is forced by the creasing blade. If the channel is too wide the crease will be soft and ineffective, while too narrow a channel will result in a pinched fold which will have small breaks and buckles along the fold line. The boards should be carefully chamfered away from the channel in order to avoid marking the sheet at the outer edges of the board.

Cutting. In order to cut completely through the sheet a sharp rule is required and extremely accurate pressure adjustment applied so that the material is cut through without damaging the edge of the rule or the impressing surface. To avoid the sheet falling apart before it is required to do so, minute nicks are made in the rule so that the sheet hangs together but can be separated readily when required.

Perforating rules are similar to cutting rules except that they are segmented to provide the 'broken cut' which is the practical effect of a perforation. Treatment is similar to cutting except that it may not be necessary to penetrate completely through the stock, to enable the sheet to remain sufficiently intact until a stress is applied. Several patterns of perforating rules are obtainable and the ratio between 'cut' and 'tie' widths will be determined by the type of material to be perforated and the degree of ease or difficulty required when the sheet is to be separated.

With all the above operations a firm counterforce is essential. The conventional paper dressing or blanket is not suitable. A metal plate or jacket is therefore fitted to avoid damaging the platen or cylinder. For simple shapes such as a centre fold for a menu or a straight perforation for a tear-off stub, the appropriate rules can be locked into a forme by a compositor or machine printer. More complicated shapes such as a folding carton, jig-saw puzzles and press out shapes for children's toys will require specialised preparation

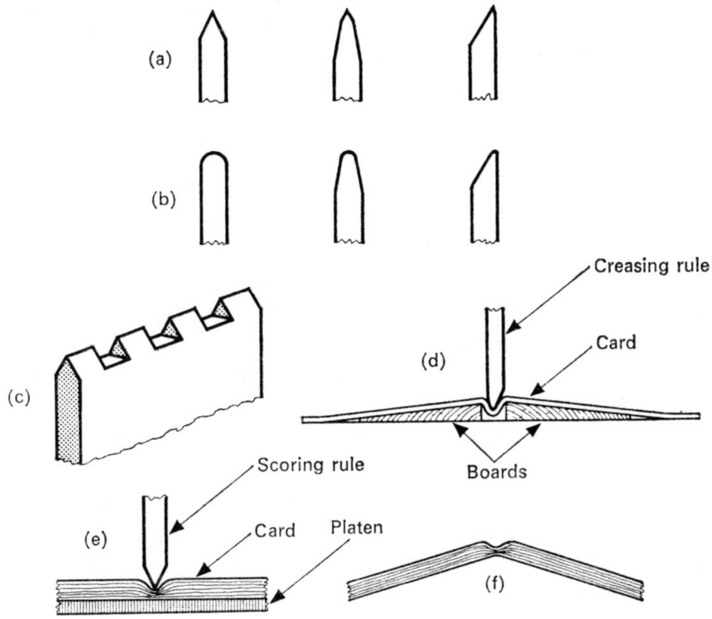

7.15 *Cutting and creasing.* (*a*) *Three types of cutting rules; centre bevel for normal work, long centre bevel for deep cuts and side bevel for close fitting.* (*b*) *Three types of creasing rules; rounded for normal work, long for deep or narrow creases and side for close fitting.* (*c*) *Perforating rules may be either centre bevelled, as shown, or side bevelled.* (*d*) *Creasing. A channel must be made for the creasing rule to enter so that the correct hinge shape is formed in the card.* (*e*) *Scoring can be done with a sharp rule similar to a cutting rule, but using less pressure so that the card is not cut through.* (*f*) *The effect of a score is to weaken the card at the fold.*

by a skilled die maker. Processing is then properly the work of the print finishing department using specialised cutting and creasing machinery, but where these machines are not available a flatbed letterpress machine may still be adapted or pressed into service. This type of forme will require more careful preparation as it may contain a combination of different types of rule, of slightly differing heights according to their purposes, each of which will require its own specialist preparation.

The inking rollers must be removed from the machine, the platen or cylinder fitted with a steel plate or jacket and a light 'impression' of the forme taken. According to the proportion of creasing rules in the forme individual channels may be built up by glueing strips of board on the jacket to accommodate the rules, or a complete board can be glued to the jacket and the channels cut out from this, remembering that channels are not necessary for other rules. A further complication is the need to push the sheet away from the cutting forme after impression, for which purpose sponge rubber pieces or small ejector springs are placed at suitable positions in the forme to ensure rapid separation.

A complicated cutting forme may be more difficult to produce to fine tolerance than the printing image with which it is to mate. For this reason it is good practice to produce the cutting forme for a carton at an early stage in planning so that the printing image may be fitted to it instead of vice-versa.

Web-fed production enables the cutting and creasing to be done in line with the printing operations as well as offering design advantages and economies due to the continuous web and elimination of the gripper margin. There are two basic types of web-fed cutting and creasing units; platen and rotary. The platen system requires the web to be fed intermittently as the platen opens and closes, and this will entail the same type of feed as described for intermittent web feed on printing machines. Formes for web-fed platen systems will not differ greatly in construction from those for sheet-fed work.

Rotary cutting and creasing, however, requires a completely different approach, as the rules must be located on a cutting cylinder. This is an expensive system demanding close engineering tolerances and careful planning to ensure that the curved rules on the cutting cylinder mate accurately with the printed design on the web. It follows that this method is only practicable for long runs where the additional work and expense will be justified by the higher speeds and better efficiency obtained when the job is running. The work from web-fed machines can either be delivered as large sheets from which the shapes can be pressed out at a later stage or the waste areas may be stripped from the web during production and the completed blanks delivered.

Appendix

SCREEN PROCESS PRINTING

Almost any type of material can be printed by this method, including the decoration of formed metal containers and inflated plastic bottles. A reverse image stencil is formed on a carrier mesh, and a flexible blade squeezes ink through the mesh to the substrate from a working puddle. The system allows for the deposit of heavy ink films, a factor of importance for printing fluorescent inks, etch resists for printed circuitry, or the decoration of nameplates etc. Screens are produced in different mesh counts and can be plain taffeta, half gauze or full gauze. Silk was originally used for mesh material but suffers fibrous interference to ink flow. Synthetic meshes such as nylon have largely taken over, and metal structures, particularly stainless steel are used for long run work.

The stencils can be drawn onto the mesh, or knife cut out of a full coating, but nowadays are widely produced by photo-development methods. Type reproduction can be achieved by using photo-composing methods without the time-consuming necessity for conversion. Halftone printing is complicated by the twin usage of screen both in the film and the process. In practice the

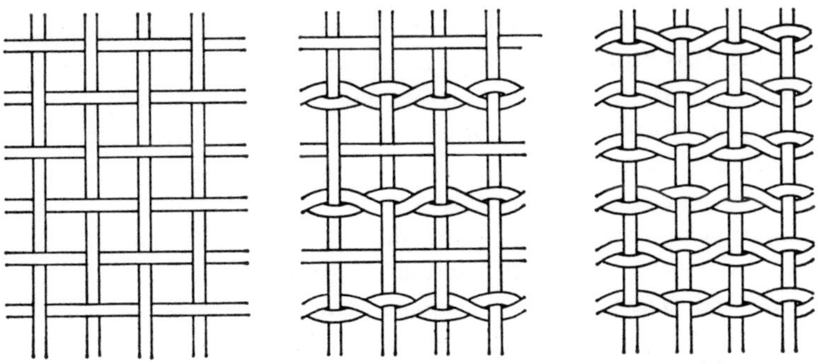

Appendix 1.1 *Types of mesh used in screen printing.* (*a*) *Plain or taffeta.* (*b*) *Half-gauze.* (*c*) *Full-gauze.*

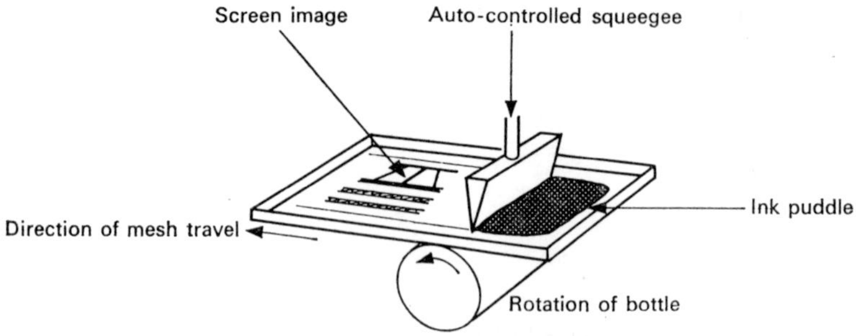

Appendix 1.2 *Bottle printing by screen process. The adaptability of screen printing can be illustrated by this example of a bottle printing device.*

dot count should be about one quarter of the mesh count or larger to give a good tonal range. The effect of the screen mesh is to lose the highlight dot while the weight of ink will tend to fill the shadow reverse dot. Allowances must therefore be made in the origination procedures.

Volatile self-solvent inks are required which will work freely, and there are specialised properties dependent upon use and the material to be printed. Since film deposits are heavy, drying requires assistance. Evaporative drying systems are used, aided by air circulation and heat. New developments have speeded the process, particularly in the field of bottle printing, and in the introduction of a web rotary system utilising internal ink ducting and squeegee control.

FLEXOGRAPHY

Formerly known as aniline printing because aniline dyes were used in the inks, flexography is a rotary process which prints from a relief rubber image with liquid inks. This process is used primarily in packaging work for such items as bags, sacks, wrappers, cartons, cases and boxes on a variety of materials. There is also a growing use of flexography outside the packaging field, particularly for books, magazines and stationery.

The advantages of flexography are ease of operation, cheap printing surfaces, quick changeovers, no makeready, and low machinery costs. Continuous designs are possible, and difficult materials such as films and foils can be printed. The prime disadvantage of flexography is its quality limitations, especially when printing halftone work, due to the use of a rubber printing surface.

The printing surface may be produced in a number of ways. The simplest is the hand-cut pattern produced from a tracing of the artist's original drawing. The patterns can be either separate plates for mounting on the plate cylinder with an adhesive, or in the form of a continuous design on a cylinder. Alternatively rubber stereos may be used. These are made from original letterpress formes using a bakelite matrix and will be synthetic rubber instead of natural rubber as used for the hand-cut plates. The rubber plates may be mounted directly on the cylinder with a liquid or dry adhesive, or they may be positioned on a flexible carrier sheet away from the machine and subsequently attached to the cylinder by a draw-bar system.

As the printing surface is made of rubber, care must be taken when formulating the inks to avoid harmful ingredients. The vehicle of the ink is usually a mixture of natural or synthetic alcohols, *eg* methylated spirits or isopropanol. Hydrocarbons such as toluol or xylol which are used for gravure are not suitable. The colouring matter is usually provided by pigments, although synthetic dyes (rarely aniline) may be used for special purposes. The third main ingredient in flexographic inks is the resins used to bind the colouring matter to the stock and to impart special properties such as gloss, flexibility and toughness. These inks dry mainly by evaporation of the volatile content, leaving behind the resin-bonded colourant on the surface of the stock. On plastic films, metallic foils and other non-absorbent stock the evaporation is usually accelerated by air blast, heating attachments, or both.

A printing unit of a flexographic machine is quite simple in construction. A rubber roller runs in a trough of the fluid ink, and by adjustment meters out a film of ink to the applicator roller. This is a steel roller with a finely-engraved pattern of cells to contain the ink and apply it directly to the printing cylinder. The printing cylinder is removable for different jobs and print-lengths and makes its impression against a plain steel cylinder from which it also obtains its mechanical drive.

Most modern flexographic printing units also have a circulating pump and filter to keep the ink clean, topped up and of even consistency. A separate motor is usually installed to drive the inking system while the machine is stationary to avoid ink drying on, and the printing cylinder should be thrown out of contact with the web and the applicator to avoid these items being welded together as the ink dries.

Flexographic printing machines are mainly web-fed although there are a few sheet-fed models available. The printing units may be built up in three basic ways to make a complete machine. They can be 1, placed in-line, which is a very useful system where very large units are used, stiff materials printed, or other in-line operations carried out; 2, stacked above each other to save space, a common type of machine for general work; or 3, the printing cylinders can be grouped around a large common impression cylinder which has

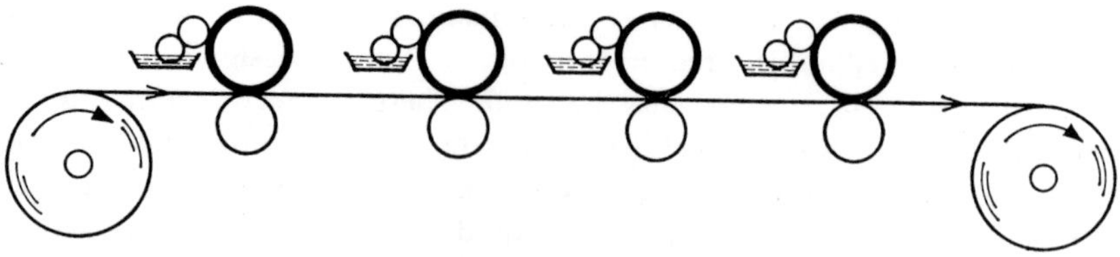

Appendix 2.1 In-line flexographic machine. This design is most suitable for large machines, rigid or semi-rigid substrates and where a number of in-line operations are also undertaken.

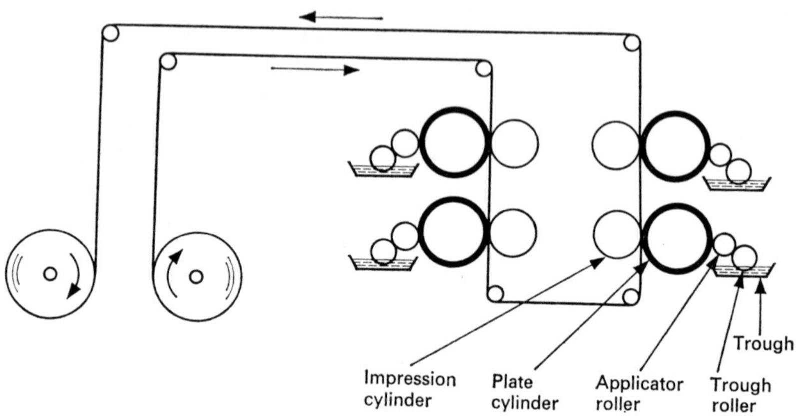

Appendix 2.2 Stack-type flexographic machine. A general purpose machine suitable for a wide range of work on a variety of packaging materials.

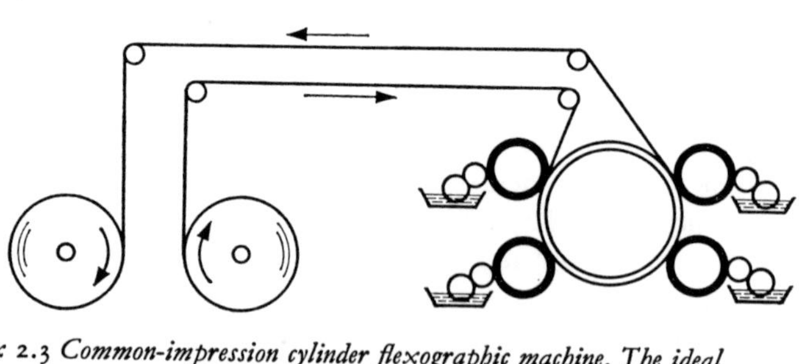

Appendix 2.3 Common-impression cylinder flexographic machine. The ideal machine for register work. The single impression cylinder ensures that even extensible materials such as polyethylene film can be satisfactorily printed.

particular advantages for printing on stretching materials and where accurate register is essential.

SOLVENTS AND ROLLER CLEANING

Roller cleaning is a task that should not be approached carelessly, or a rapid deterioration in surface quality can result. An effective solvent will reduce polarity and flocculate the pigment, but a too powerful one will increase adhesion and make dispersion more difficult when using automatic wash-up devices with mechanical cleaning action. Most letterpress and litho inks can be cleaned with hydrocarbon solvents such as paraffin, white spirit and distillates, with the principal exception of moisture-set inks. It is wrong to try to remove dried ink films from rollers by solvent action; swelling agents should be used

Properties of some common photogravure and flexographic solvents

Solvent	*Boiling range*	*Flash point (closed cup)*	*Relative evaporating time (ether = 1)*
Hydrocarbons			
SBP2	71–94°C	−18°C	1·3
SPB3	120–192°C	−8°C	2·1
Toluene	105–111°C	4°C	6·1
Xylene	136–140°C	25–28°C	13·1
Alcohols			
Industrial Methylated Spirit	77·5–79°C	13–16°C	8·0
Isopropanol	79·5–81·5°C	12°C	21·0
Methanol	64·5–66°C	0°C	6·3
Butanol	114–118°C	35°C	33·0
Glycols			
Cellosolve	130–136°C	40°C	40·0
Esters			
Ethyl Acetate	76–79°C	−5°C	28·5
Isopropyl Acetate	87–90°C	7°C	42·0
Butyl Acetate	124–128°C	25°C	11·8
Ketones			
Acetone	55·5–56·5°C	−18°C	2·1
Methyl Ethyl Ketone	79–80·5°C	−7°C	6·3
Methyl Isobutyl Ketone	114–118°C	16°C	17·0

to loosen the molecular bonds between ink and roller. On litho presses a certain amount of gum and paper detritus finds its way onto the rollers and is not readily soluble in the same solvent as ink. A two part wash-up system can aid cleaning, first with an application of solvent containing water or hygroscopic material, and a final application of evaporative solvent to clear moisture traces.

Aliphatic or 'straight chain' solvents – distillates, naptha, kerosene – are safe on all types of roller, but aromatic hydrocarbons such as benzine, toluene, xylene (xylol) etc, will swell most rubbers if used in quantity. Ketonic solvents – the oxygenated solvents such as acetone, cyclohexanone, methyl-ethyl-ketone – can be used to augment the aliphatic solvent to clean butyl-rubber, but should not be used on gelatine or polyurethane rollers. Anhydrous solvents such as alcohols and glycols must also be avoided on gelatine and polyurethane. Oxygenated solvents, esters, ethers, and chlorinated solvents all have a detrimental effect on roller surfaces proportional to time in contact, and should be reserved for special cleaning problems only. The use of any but

Comparative solvent resistance of roller coverings

Solvent	*Gelatine*	*Poly-urethane*	*Natural rubber*	*Neoprene*	*Nitrile*	*Butyl*	*Poly-sulphide*
Animal and vegetable oils	Excellent	Excellent	Poor	Good	Excellent	Good	Very good
Aliphatic solvents	Excellent	Excellent	Poor	Good	Very good	Fair	Good
Aromatic solvents	Excellent	Fair	Poor	Poor	Fair	Poor	Good
Alcohols & glycols	Poor	Fair	Good	Excellent	Excellent	Good	Very good
Ketones and acetates	Fair	Very poor	Fair	Poor	Very poor	Good	Good
Esters	Good	Poor	Poor	Fair	Poor	Poor	Fair
Ethers	Fair	Poor	Poor	Poor	Poor	Good	Fair
Chlorinated solvents	Good	Very poor	Very poor	Poor	Poor	Fair	Good

aliphatic or oil solvents on polyurethane gradually causes the roller surface to soften and lose shape and consequently inks should not be allowed to dry on them.

The automatic wash-up devices are generally of two basic types. One employs an essentially soft rubber or leather blade set at a near perpendicular angle to either scoop or trail; the second type utilises a harder plastic or metal blade set at a sharp tangent to the wash-up drum to scoop more effectively. The second system gives a faster wash-up but may be more wearing on the roller being doctored. Modern designs employ surfaces giving minimal friction problems consistent with maximum contact.

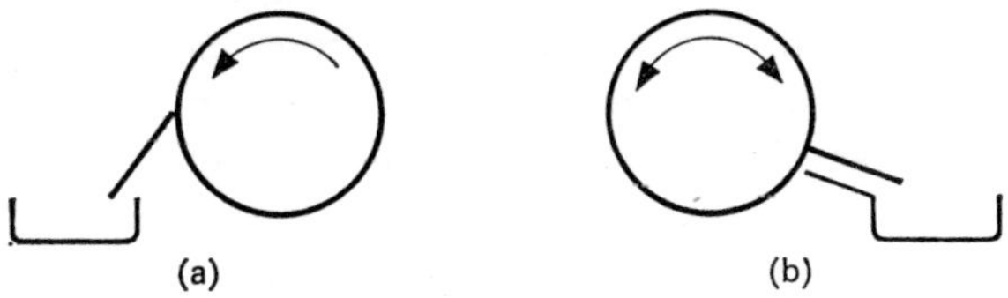

Appendix 4.1 *Automatic washup units.* (*a*) *Scoop only; hard doctor blade at a sharp angle.* (*b*) *Can scoop or trail; soft doctor blade at a near perpendicular angle.*

SOLVENT RECOVERY

In general, gravure inks have from 50 to 70 per cent solvent content. Many of these solvents are expensive and once they have conveyed their ingredients to a substrate are removed by air turbulence and extraction, then dispersed into the outer atmosphere, subject to pollution regulations, or recovered.

To obtain the maximum efficiency from the recovery plant the design of the drying hoods on the press must ensure the removal and extraction of the maximum amount of solvent from the ink film. The solvent laden air is taken along air ducts to the recovery plant where it is pumped into the recovery chamber. Warm air is pumped in beneath it and as the warm air rises through the chamber, the centre of which is filled with carbon pellets, the solvent is extracted from the air by the carbon and the free air is allowed to pass out through a valve at the top of the adsorber. When the carbon is completely saturated with solvent, the flow of air is stopped and steam is then passed through the chamber from the top. As the steam passes through the carbon it takes the solvent with it and both pass into a condenser which turns the steam back to water. A separator unit parts the two liquids and passes the solvent to a storage tank.

Solvent recovery plants are built with efficiency, safety, reliability and the minimum of operating personnel and maintenance in mind. The efficiency of a recovery plant is estimated at 97 per cent. There are two types: 1, the static recovery method, and 2, the continuous recovery plant.

The static recovery plant is built to meet the needs of the customer and will vary in shape, size and capacity. It usually consists of two large horizontal cylindrical adsorbers built on to a casting unit. The adsorbers are bisected by monel metal gauzes which support the acticarbon extruded pellets. Carbon is used because of high absorption efficiency and to ensure freedom from dust, and depending on size as much as 6 tonnes may be used in each adsorber. In order to ensure safety the flow of solvent laden air is held at a pre-determined figure and in the event of a failure causing the volume of air to fall below a safe minimum, valves cause large steam ejectors to operate and restore the extraction flow. At the same time the plant shuts down and can only be restarted manually. If the steam pressure falls below a safe minimum the whole plant shuts down and can only be restarted when the pressure is restored. In order to avoid leaving the plant shut down with a heated wet carbon mass, a special shut down cycle is started in which air is blown through for a period of time to dry the carbon pellets.

The continuous recovery plant has a rotating adsorber housed in an enclosed chamber. Solvent laden air is ducted to the plant, passed through a cooling unit and from thence into the chamber. The rotary adsorber is divided into a

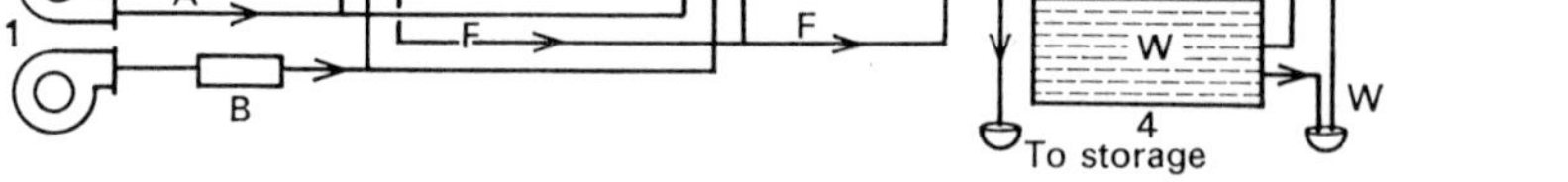

A. Solvent laden air
B. Heat exchanger
C. Carbon
D. Free air

E. Steam
F. Distillate
H. Cooling water
S. Solvent

W. Water
FWH Factory, washing, heating

1. Blower fans
2. Absorbers

3. Condenser
4. Tank

Appendix 5.1 *Continuous solvent recovery plant.*

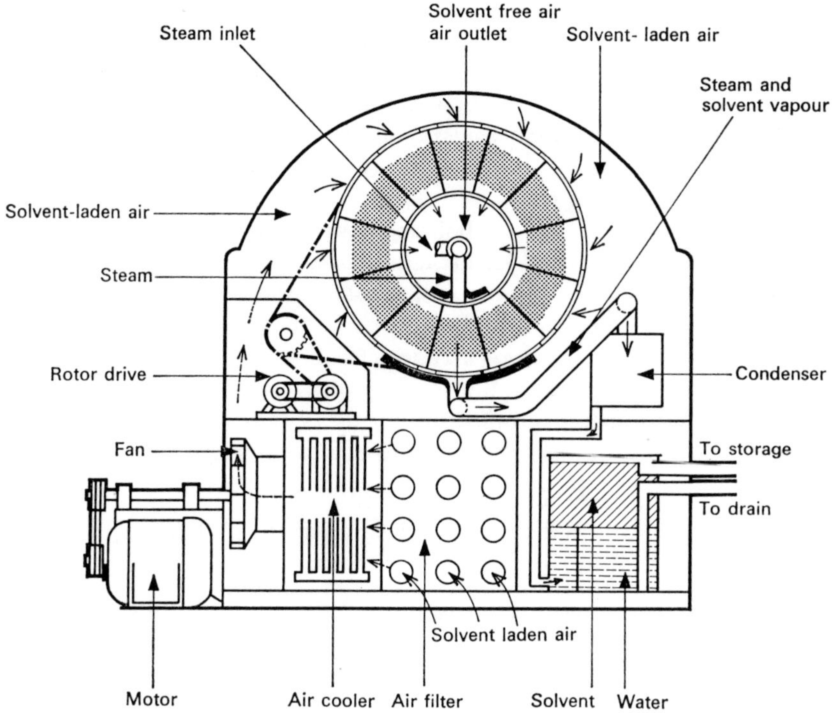

Appendix 5.2 Rotary bed solvent recovery unit.

number of sections, each housing carbon pellets. Steam is introduced into the chamber where steam and solvent laden air pass through the carbon filters. The free air escapes from the centre of the adsorber and the steam and solvent mixture are ducted from below the centre of the adsorber to a condenser unit. The condenser unit changes the steam to water; water and solvent separate and solvent is piped to a storage tank while the water is drained away.

RELEVANT BRITISH STANDARDS

BS1480:1949 Letterpress process inks.
BS2489:1954 Sequence of measurements for printed matter.
BS2650:1955 Four-colour offset lithographic inks.
BS2913:1957 Measurement of offset blanket thickness for lithographic printing.
BS2984:1958 Letterpress printing plates. Part 1: duplicate plates.

BS3020:1959 Improved inks for letterpress four-colour printing.
BS3319:1960 Type-high numbering boxes.
BS3649:1963 Original printing plates.
BS3763:1964 The International System (SI) units.
BS3814:1964 Glossary of letterpress rotary printing terms.
BS 950:1966 Viewing conditions in the graphic arts.
BS1133:1967 Packaging code.
BS1991:1967 Letters, signs, symbols and abbreviations.
BS4149:1967 Glossary of paper/ink terms for letterpress printing.
BS4160:1967 Inks for letterpress 3- or 4-colour printing.
BS4000:1968 Sizes of paper and board.
BS4277:1968 Glossary of terms used in offset lithographic printing.
BS4321:1969 Methods of test for printing inks.
PD5686:1969 The use of SI units.
BS4666:1971 Inks for offset three or four-colour printing.

The above publications are available from:
British Standards Institution, 101/103 Pentonville Road, London, N1

Glossary

Air knife: stream of air in a drying system, forced through a narrow slit at high pressure.

Arrow (comet): printing fault in gravure which appears as a mark in the shape of an arrow head on the print.

Back up roller (bowl): roller which backs up the impression roller for additional pressure in gravure printing.

Baffles: metal plates in a drying device which control air turbulence.

Bar mark (kick): printing fault in gravure, showing as a variation in pressure across the cylinder.

Bay window: arrangement in a reel-fed machine allowing a strip of the web to be taken out of the main run and re-inserted at a different position.

Behringer system: method of ink application on gravure machines employing a primary blade to pre-meter the ink before doctoring.

Bi-metal plate: lithographic printing plate with two differently-selective metal surfaces.

Binder: adhesive agent in an ink, usually a varnish or resin.

Bleed: 1, image areas designed to print over the edge of a sheet. 2, migration of part of an ink through the substrate after printing.

Block: letterpress printing plate on its own individual base.

Blocking: sticking together of successive turns of material in a rewind, usually a result of incomplete drying.

Bowl roller (back up): roller which backs up the impression roller for additional pressure in gravure printing.

Butterfly: printing fault in gravure appearing on the print in the shape of a pair of wings.

Catching up: lithographic term for ink transfer to the non-image areas of the printing plate.

Colour bars: strips of solid and tint colour printed in the margins of a job to aid the control of hue, density and ink film thickness.

Colour halftone: translation of colour photographs into dot formations to simulate the original copy.

Comet (arrow): printing fault in gravure which appears as a mark in the shape of an arrow head on the print.

Cow-catcher: auxiliary pile support to collect delivered printed sheets during pile changeovers. The name derives from the fork-shaped construction of early models.

Crawling (reticulation): printability fault in which ink gathers together into small droplets due to complete incompatability with the substrate.

Dahlgren: press engineer who converted the basic principle of applying moisture to a lithographic plate via the inking system so that it could be applicable to high-speed large offset machines.

Decalcomania: printed substrate in which the image is to be transferred to another substrate (such as pottery).

Diazo: term recognising the two nitrogen atoms which characterise diazonium salts. Diazo materials form the basis for most pre-sensitised platemaking systems.

Doctor: blade which shears the ink from a photogravure printing surface.

Doughnutting: photogravure printing fault in which ink does not transfer from the centres of the cells.

Dressing: covering material on an impression cylinder.

Duct: ink container in letterpress and litho designed to meter out a film of ink.

Durometer: instrument for measuring the hardness of a resilient material.

Festoon: series of loops in a paper web.

Filling-in: printing fault in which ink clogs up small open spaces.

Flash point: temperature at which a concentration of solvent vapours will ignite if exposed to a flame or spark.

Flow cup: open cup with a lower orifice of specific diameter, used as a control device for liquid ink viscosity.

Fluffing: effect of loose-surfaced papers being pulled by the ink on a printing surface.

Fountain: container for the damping solution on a lithographic machine.

Fountain solution: aqueous solution applied to a lithographic printing plate.

Gear mark: fault appearing as lines across the printed image related to the pitch of the gears.

Ghosting: effect due to the printed image on one side of a sheet affecting the drying of the print on the other side.

Grain: 1, direction in which fibres tend to lay in a sheet or web of paper. Also known as machine direction. 2, roughened condition of a lithographic printing surface which enables it to retain moisture.

Guttering: fault in letterpress printing in which the impression cylinder drops in the spaces between image areas.

Halftone: reproduction of a continuous tone image by interpreting it as a pattern of dots of varying sizes.

Halo: 1, ridge of ink around a letterpress printed image caused by over-impression or over-inking. 2, spreading of ink through a substrate after printing resulting in a diffused boundary to the image edges.

Hard flame: gas flame under high pressure, impinging on the printed web in a drying system.

Heat sealing: joining a material to itself or another by the application of heat and pressure.

Hickey: particle of dried ink or paper detritus on a printing plate or offset blanket causing spots to appear on the print.

Idler (pipe roller): free-running roller on a reel-fed press which helps to guide the web through the machine.

Impression nip: contact area between the printing and impression surfaces.

Interlay: sheets of paper or other material of varying thickness placed between the plate and its base in a letterpress forme to provide corrected pressure to the various tones.

Inverted halftone: gravure image in which the cells vary in area according to the tone values required.

K and N ink: proprietary test ink of standardised rheological characteristics and colour, used to test the relative absorbency of paper surfaces.

Key forme: printing forme which is used to determine register for the other workings.

Key cylinder: cylinder which is used to determine register for the other cylinders on a reel-fed machine.

Key sheet: pre-printed master sheet used for registering purposes.

Kick (bar mark): variation in print strength across the cylinder in gravure, caused by a change in the doctor blade pressure.

Knocking up: alignment of all the sheets in a stack of paper to ensure accurate positioning for feeding and registering.

Lay: precise position to which sheets of paper are laid to ensure accurate registration.

Lazy flame: 'soft' gas flame under low pressure, which follows the web for a short distance in a drying system.

Machine direction: direction in which the fibres tend to lay in a sheet or web of paper. Also known as grain direction.

Makeready: 1, operations necessary to prepare a job for printing. 2, in letterpress, adjustment of printing pressure according to tonal requirements.

Medium: colourless printing ink used as an extender in gravure printing.

Metamerism: effect in which a colour viewed by different light sources appears to vary in hue.

Misting: haze caused by ink droplets in the air around printing units.

Moiré: patterning effect in colour halftone printing caused by a clash of two adjacent screens.

Molleton: thick cotton fabric, similar to flannel and having a long nap. Used for covering damping rollers in lithography.

Mottle: printing fault in which solid areas have a dappled appearance.

Multi-metal: lithographic printing plate comprising several metals, two of which are used for the image and non-image areas due to their selective properties.

Nip: contact area between two adjacent surfaces.

Oscillator: roller in an inking system which moves laterally across the machine in order to break down the structure of the ink.

Overlay: sheets of paper or other material of varying thickness placed in position on the impressing surface of a letterpress machine to provide corrected pressures to the various tones.

pH: measure of the hydrogen ion dissociation in an aqueous solution, indicating its degree of acidity or alkalinity.

Picking: printing fault in which the surface of a coated stock is pulled away by the ink.

Pin bar: device for locating register work, employing a set of pins or studs on a common bar.

Pinholing: printing fault in which small holes can be detected in otherwise solid inked areas.

Pipe roller (idler): free-turning roller on a reel-fed press which helps to guide the web through the machine.

PIV: (positive infinitely-variable) gearing system which provides smooth, stepless speed variations.

Powdering: printing fault in which the ink dries to a powder on the surface of the substrate.

Pre-tension: stress deliberately introduced between a pair of cylinders to ensure their contact under printing pressure.

Printability: compatability of ink and paper.

Process: colour halftone printing.

Progressive proofs: set of colour halftone proofs produced by the platemaker as a guide for machine printing.

Proof: printed sheet produced prior to the machine run for customer approval, checking or control purposes.

Proving: taking proofs for customer approval, checking or control purposes.

Pulling groups (feed rollers): driven rollers which control the feed of paper in a reel-fed machine.

Quoin: expanding device included in a letterpress forme to lock the various elements into the chase.

Refusal: failure of one ink film to print upon another.

Register: relative positions of two or more items.

Register roller: roller in the path of a web which can be adjusted to alter the length of web between two operations.

Repeat: printing fault associated with the inking system in which the rollers are not fully replenished while rolling the image, resulting in inadequate coverage of parts of the forme.

Reproduction pull: print, usually of type, taken for reproduction by a photographic process.

Reticulation (crawling): printability fault in which ink gathers together in small droplets due to complete incompatability with the substrate.

Reverse pyramid: gravure image in which the cells are tapered in depth.

Satellite: web offset printing unit designed on the common impression cylinder principle.

Screen clash: patterning effect caused by the interference of two superimposed screens.

Scumming: lithographic ink printing as a fine tone in non-printing areas.

Selsyn motor: small, self-synchronising motor used for transmitting signals.

Serrated roller: roller with a knurled or milled surface to reduce ink pick-up and to obtain better web traction on reel-fed machines.

Set-off: unwanted transfer of ink from a print to another surface.

Shore hardness: measure of the degree of compressibility of resilient materials where 0 = air and 100 – glass hardness. Used on the Durometer (q.v.).

Show through: degree to which printed matter can be seen through its substrate.

Slur: printing fault caused by a false movement between the inked image and the substrate.

Small offset: term originally referring to offset office duplicators, but now extended to include machines of lightweight construction and simplified operation printing sheets up to A2 size.

Splice: attaching a new reel to an expiring web on a reel-fed machine.

Split-press work: using a number of in-line printing units so that the product from certain units will run into a separate folder from the others.

Spot colour: printing colour in local areas or patches, as distinct from colour halftone work.

Stability index: measure of the amount of pre-tension (q.v.) in a printing unit.

Step wedge: quality control device included with the printing image comprising a series of neutral density patches.

Stock: 1, substrate to be printed upon; usually applied to paper and board materials. 2, spindle of a covered printing roller.

Streaking: printing fault which appears as lines of varying density on the print.

Stream feeder: method of feeding sheets into a machine in a continuous flow; the sheets overlapping each other as they travel to the machine.

Strike through: printability fault in which ink penetrates completely through a porous substrate to appear on the reverse side.

Stroboscope: device, employing intermittent illumination, to study high-speed operations with regular, periodic movement.

Substrate: material which is printed upon.

Swing grippers: set of grippers which take a sheet from the feedboard and accelerate it to meet the grippers on the impression cylinder. Also known as transfer grippers.

Tack: stickiness of an ink.

Thixotropy: property shown by certain semi-liquids, such as printing inks, in which a structured body is formed when standing for a period, but is broken down when shaken or stirred.

Tone band: gravure printing fault showing as lines on the printed sheet, caused by foreign matter lodging behind the doctor.

Toner: dye used to modify the colour of a printing ink.

Transfer grippers: set of grippers which take a sheet from the feedboard and accelerate it to meet the grippers on the impression cylinder. Also known as swing grippers.

Trapping: correct laying of one wet ink film upon another.

Trellis: pattern formed on a gravure print, caused by foreign matter lodging under the doctor blade.

Trip: control device on a printing machine to bring an operation into sequence.

Tumble (work and tumble): turning a sheet over to the opposite long edge for perfecting.

Turn (work and turn): turning a sheet over for perfecting, using the same front lay edge.

Twist (work and twist): turning a sheet through 90° for a second printing on the same side.

Twisting rollers: pair of rollers either side of a printing unit on a reel-fed machine, which allow a limited adjustment of helical register.

Undercolour removal: removal of unnecessary colour in black or shaded areas of a set of colour halftones.

Underlay: bringing up to correct height of a letterpress block by means of packing under the base.

Vehicle: film-forming constituent of an ink which gives it 'body' and binds the colourant to the substrate.

Web: material unwound from a reel.

Web tension: stress applied to a web in a machine for control purposes.

Wet-on-wet: printing subsequent films of ink upon each other before the first has dried.

Wettability: degree to which a liquid will spread upon a surface.

Wipe: printing fault caused by a false movement between the inking rollers and the forme.

Index

Index

Index